GEOMETRY

ESSENTIAL REVIEW FOR AP®, HONORS,
AND OTHER ADVANCED STUDY

By the High School Experts at
THE PRINCETON REVIEW

Penguin
Random
House

The Princeton Review
110 East 42nd St, 7th Floor
New York, NY 10017
Email: editorialsupport@review.com

Published in the United States by Penguin Random House LLC, New York, and in Canada by Random House of Canada, a division of Penguin Random House Ltd., Toronto.

Some of the content in Fast Track: Geometry has previously appeared in Basic Skills for the GED® Test, published as a trade paperback by Random House, an imprint and division of Penguin Random House LLC, in 2017

This book was previously published in a different format with the title High School Unlocked: Geometry by the Princeton Review, an imprint of Penguin Random House LLC, in 2017.

ISBN: 978-0-525-57172-8
eBook ISBN: 978-0-525-57186-5
ISSN: 2767-0910

Editor: Chris Chimera
Production Editors:
Sarah Litt and Kathy Carter
Production Artist: Gabriel Berlin
Content Contributor: Heidi Torres

Printed in China.

10 9 8 7 6 5 4 3 2 1

For customer service, please contact **editorialsupport@review.com,** and be sure to include:
- full title of the book
- ISBN
- page number

ACKNOWLEDGMENTS

The editor would like to thank Heidi Torres for her dedication and expertise to her wonderwork for this edition.

The editor would also like to thank Gabriel Berlin, Sarah Litt, and Kathy Carter for their careful attention to detail on each page of this book.

CONTENTS

WHERE TO GET MORE (FREE) CONTENT

Scan the QR Code:

OR

Go to PrincetonReview.com/prep

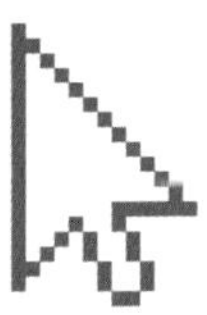

THEN:

- Enter the following ISBN for your book: 9780525571728
- Answer a few simple questions to set up an exclusive Princeton
- Click the "Student Tools" button, also found under "My Account" from the top toolbar. You're all set to access your bonus content!

Need to report a potential content issue?

Contact EditorialSupport@review.com and include:

- full title of the book
- ISBN number
- page number

Need to report a technical issue?

Contact TPRStudentTech@review.com and provide:

- your full name
- email address used to register the book
- full book title and ISBN
- computer OS (Mac/PC) and browser (Firefox, Safari, etc.)

Once you've registered, you can...

- Get valuable advice about the college application process, including tips for writing a great essay and where to apply for financial aid
- Check to see if there are any corrections or updates to this edition
- If you're still choosing between colleges, use our searchable rankings of the latest edition of The Best Colleges to find out more information about your dream school.

INTRODUCTION

WHAT IS THIS BOOK AND WHEN SHOULD I USE IT?

Fast Track: Geometry is a one-stop guide meant to help you keep up with your course—or to accelerate your content expertise. Whether you're a crammer, visual learner, or high-level student looking for extra review, this book is filled to the brim with information that you'll benefit from knowing (or brushing up on). We've taken every opportunity to translate key content into friendly and succinct formats, with annotated illustrations showing every part of important theorems of geometry and much more. In short—which is the point of this book—think of the following material as a fantastic set of class notes.

This book is pocket-sized because it's meant to be used whenever and wherever you want. Waiting on a late friend? Spend some time with quadrilaterals! Have some time to kill before practice? Warm up with a little trigonometry refresh! These are bite-sized topics that you can pick up and put down at your own pace. Whether you need it for a last-minute review or as a supplement for your class, you'll find no judgment here. Remember: it's never too late (or too early) to start studying.

THIS ISN'T STANDARDIZED TEST PREP

What you won't find here are test-taking strategies or practice questions for a specific standardized test, like the AP, ACT, or SAT; for that kind of focus, check out our test-specific Math titles online or at your local bookstore.

WHAT DOES THIS BOOK COVER?

The start of this book takes you all the way back to the fundamental building blocks of all things: vocab. We will explore key terms from geometry in depth, and pay attention to that first chapter, because everything builds from there. Once we have the foundation laid, we will explore properties of triangles, quadrilaterals, polygons, and circles. We will review congruence and similarity, continuing on to constructions and transformations. Then we will progress to three-dimensional figures. Finally, we cover how to do geometric proofs.

In short, this book covers all of the geometry content—and probably then some—that you would study in a high school geometry class! This means that whether you're cramming for your class's final exam or prepping for an AP or SAT-level test in geometry, this book can help you get there—on the fast track.

HOW DO I USE THIS BOOK?

Think of this book like a toolkit: figure out what you're trying to accomplish, and then choose the part that works best for your schedule and your needs.

DIG DEEP

Start by looking at the **ASK YOURSELF** questions throughout these pages. You're checking to see where your knowledge might be a bit loose, testing your connections, and identifying tough spots that may need special attention.

OPEN THE BRAIN VALVES

Once you've identified a trouble spot for yourself, target that specific chapter (or section). Sometimes you just need to replace some bad connections with some new ones, and these visualizations will help you get the job done!

NAIL THAT INFORMATION INTO PLACE

Pick topics at random and try to explain or summarize them. Can you reproduce notes on the subject? Really make sure you can visualize each concept: for instance, when you think about the area of a trapezoid, can you see the triangles the trapezoid can break into to derive the formula?

DO IT ALL

There's no wrong way to use this book! Read it cover to cover—more than once if that's what helps you to connect each piece. Try focusing on the graphics the first time through, and then go back through and line them up with the text—or read the text first, and then use the images to solidify those thoughts.

WHAT DO THE DIFFERENT SIDEBAR TYPES MEAN?

You may not realize it, but every class you take is carefully aligned to some set of standards that your instructor, principal, or even local government has determined needs to be covered. What you're getting, then, is what may have worked in the past, or what's comfortable for your instructor—which isn't necessarily what's best for you, and which may cause the class to feel overwhelming.

Instead of going at the teacher's pace, *Fast Track: Geometry* works more like a self-guided tour of the subject. If there's something you're interested in, you can linger on that topic instead of sitting back as the class glosses over it. To that end, the graphics and icons are designed to make you read more actively and to keep you engaged. Most of the visuals will be self-explanatory, but there are a few features that will pop up over and over again, which we'd like to explain here:

ASK YOURSELF...

ASK YOURSELF questions look like this. These are opportunities to solidify your understanding of the material you've just covered. They're also a great way to take these concepts outside of the book and make the sort of real-world connections you'll be able to use in any essays or short-answer questions you come across.

REMEMBER...

The **REMEMBER** symbol indicates certain facts that are likely to come up again in different sections, or important information that later sections may build on.

CHAPTER

1

INTRO AND VOCAB

© istockphoto.com / artisteer

You can't understand complex geometric concepts without knowing the basics. Completing a formal proof that a particular quadrilateral is a rhombus requires you to know the basic definition of angles, parallel lines, and even points. We'll review some of these ideas in this chapter.

CHAPTER CONTENTS

INTRODUCTORY VOCABULARY

In this chapter, we'll be reviewing some important basics for Geometry. First, a little vocab:

A **point** indicates an exact position.

• A

A point has no size and no dimensions (such as "length" or "height"). On paper, we usually represent a point with a dot. Often, a point will be labeled with a "name," usually a capital letter such as *A* or *B*. This is so that we can identify it specifically when discussing a problem.

A **line** represents a straight path, and it extends infinitely in two opposite directions.

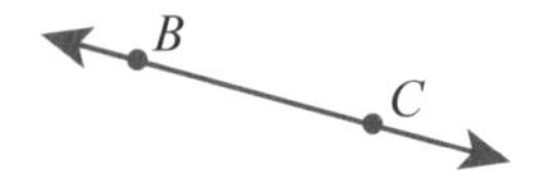

A line has one dimension (length), but it has no thickness. A line has no curves or corners. Since it extends infinitely in both directions, it has no endpoints. On paper, we usually represent a line using arrows on each end. We can "name" a line using its points (such as line *BC* above), or with another label, usually a lowercase letter (such as line ℓ).

Through any two points, there exists exactly one line. In other words, it takes two points to define a unique line.

A line contains infinitely many points. That doesn't mean that it contains all possible points that exist in the surrounding space—it contains only the points that lie on the path of that line.

A **plane** represents a flat space that extends infinitely in two dimensions.

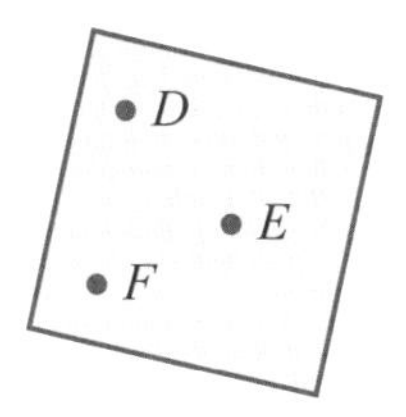

A plane has two dimensions (length and width), but no thickness. We can name a plane using its points (such as plane *DEF* above), or with another label, usually a capital letter (e.g., plane *P*).

When we talk about a two-dimensional shape, such as a circle or square, the shape is considered to lie "on a plane," meaning that it is flat.

A plane contains infinitely many points and infinitely many lines.

Through any three points, there is exactly one plane—as long as the three points don't lie on the same line. That is, it takes at least three points to define a unique plane.

Points that lie on the same line are **collinear** points. Points that lie in the same plane are **coplanar** points.

Space is the term for all possible points in three dimensions.

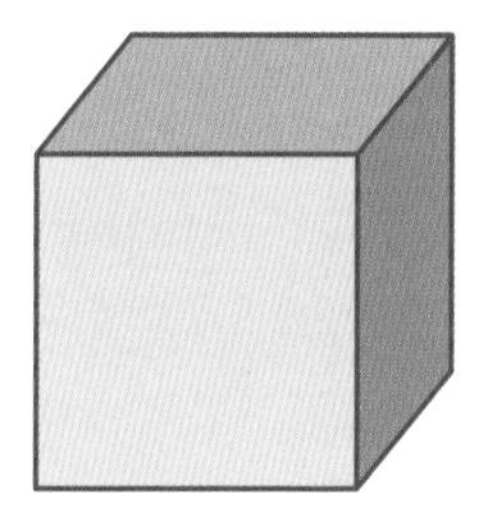

A three-dimensional shape (such as a sphere or cube) is called a **solid**. Spaces and solids have three dimensions—length, width, and height.

A **segment** is part of a line and has two endpoints.

We name a segment using its two endpoints (such as segment *AB* above). A segment contains infinitely many points—all between its two endpoints.

A **ray** is part of a line. It has one endpoint, and extends infinitely in one direction.

We usually represent a ray with a dot at the endpoint and an arrow at the other end. We can name a ray using two of its points (such as ray *CD* above). A ray has infinitely many points.

In geometry, **congruent** means *having exactly the same shape and size.*

- Congruent line segments have exactly the same length.
- Congruent angles have exactly the same degree measure.
- Congruent shapes have the same side lengths and angle measures.

In text, congruency is indicated by the congruent sign (≅). In figures, congruency is indicated by hash marks. For example

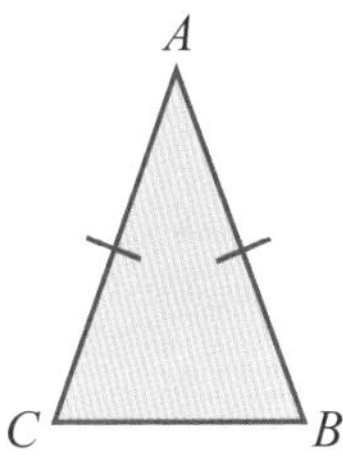

In the figure above, $\overline{AB} \cong \overline{BC}$.

ANGLES

An **angle** is formed by two rays with the same endpoint.

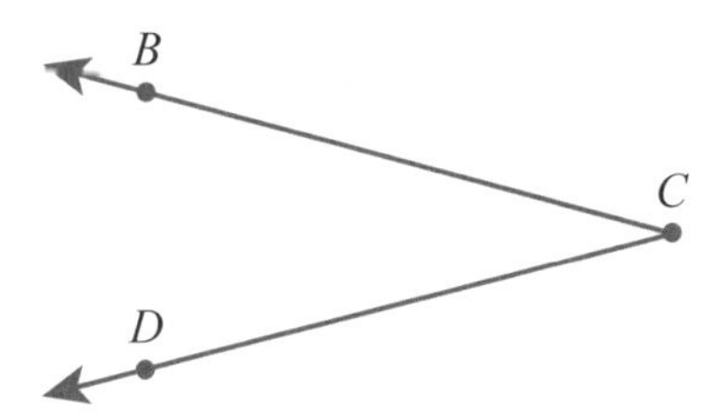

The shared endpoint is the **vertex** of the angle. (Note: the plural of vertex is **vertices.**) We usually name an angle by its vertex (such as angle *C*, above) or with three points (such as angle *BCD*).

Angles are usually measured in **degrees** (represented by the symbol °). The measure tells us the "size" of the angle, or how much it opens.

TYPES OF ANGLES

An **acute** angle is between 0° and 90°.

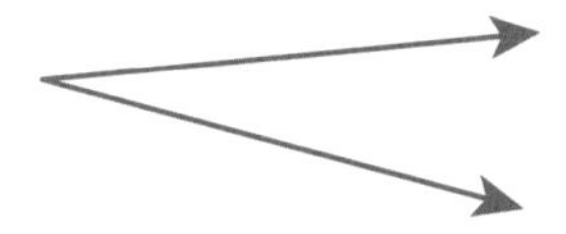

An **obtuse** angle is between 90° and 180°.

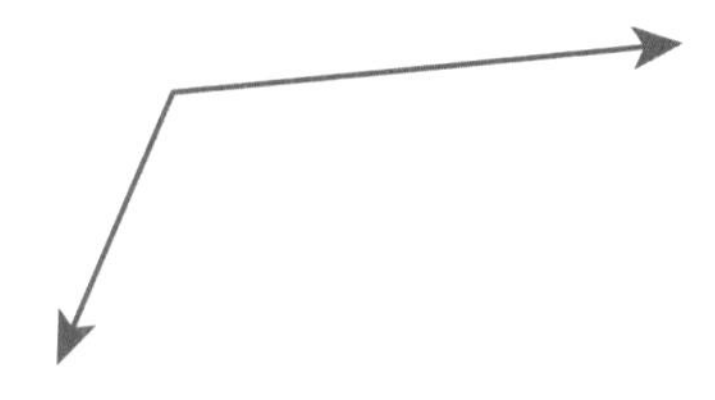

A **right** angle is exactly 90°. Right angles are often represented with a small square symbol at the vertex.

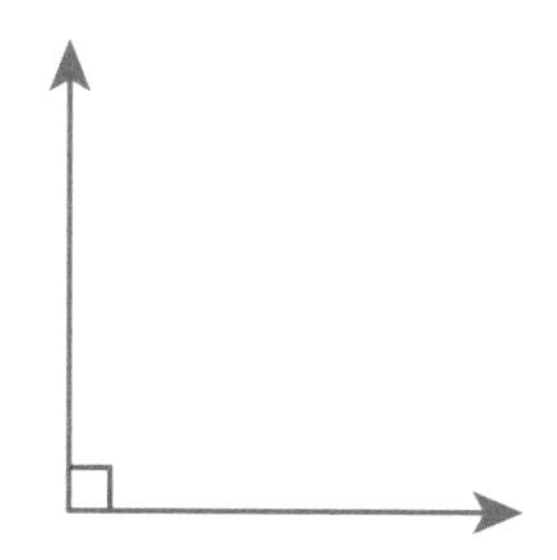

A **straight** angle is exactly 180°. A straight angle forms a line.

ANGLE PAIRS

When referring to pairs of angles, there are certain conditions that have special names. Knowing about angle pairs is useful for discussing shapes and solving for unknown angle measures.

Congruent angles are angles with the same measure. In geometry, congruent more generally means "having the same exact shape and size."

Complementary angles are two angles whose measures add up to 90°. In the figure below, A and B are complementary angles. You can also say that A and B are **complements** of each other.

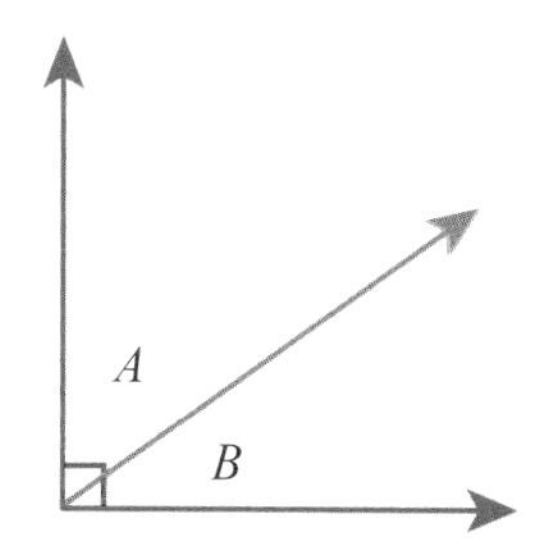

Supplementary angles are two angles whose measures add up to 180°. In the figure below, *C* and *D* are supplementary angles. You can also say that *C* and *D* are **supplements** of each other.

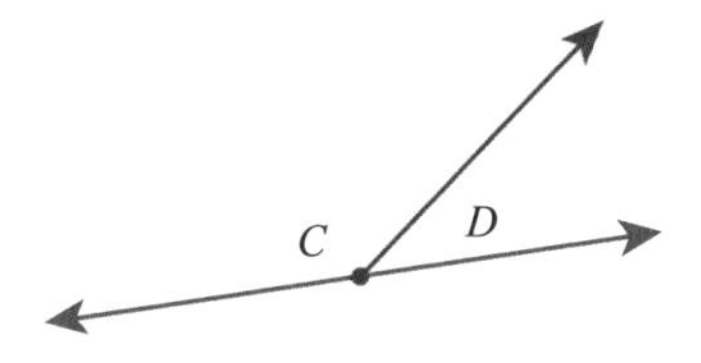

Another term for supplementary angles is **linear pair**, since together they form a line.

Adjacent angles are two angles that share a common vertex and a common side. (In everyday language, adjacent means "next to or adjoining.") The common vertex and common side are shown in blue in the figure below.

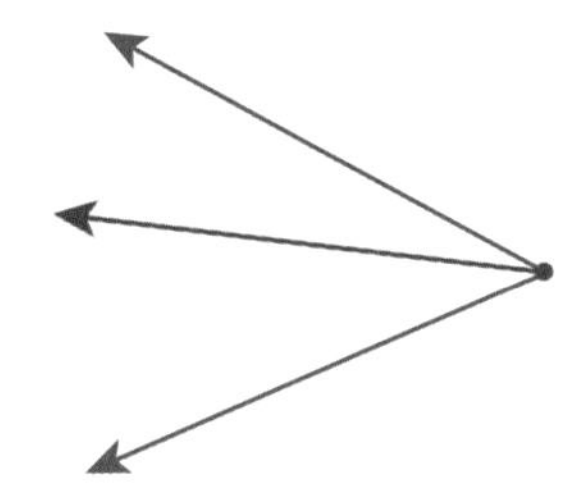

Vertical angles also share a common vertex, and are formed from two intersecting lines, as shown. Vertical angles have equal measure.

The term *vertical angles* refers to the shared *vertex*. It's not related to the usual meaning of vertical, as in "up and down."

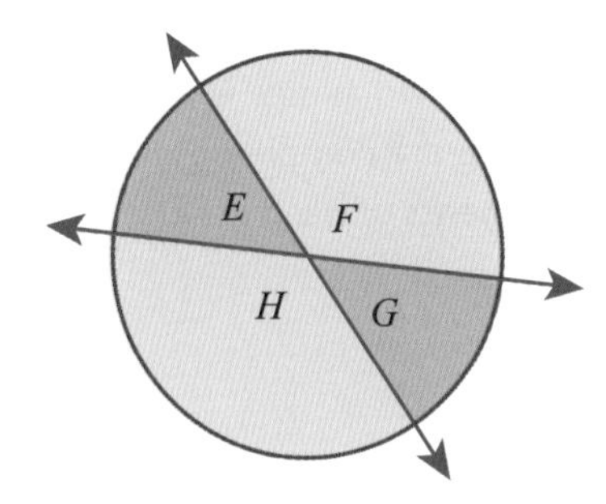

The *vertical angles* are the angles opposite each other. In the diagram above, *E* and *G* are vertical angles, and *F* and *H* are vertical angles. Whenever two lines intersect, two pairs of vertical angles are formed.

PARALLEL AND PERPENDICULAR LINES

Parallel lines are lines that never intersect. In other words, the lines have a constant distance between them. The symbol || means parallel.

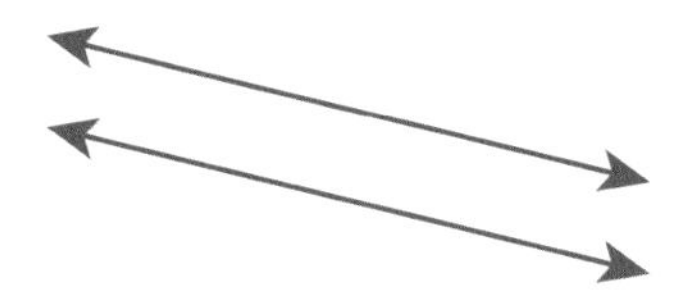

Perpendicular lines are lines that intersect at a right angle (90°). The symbol ⊥ means perpendicular.

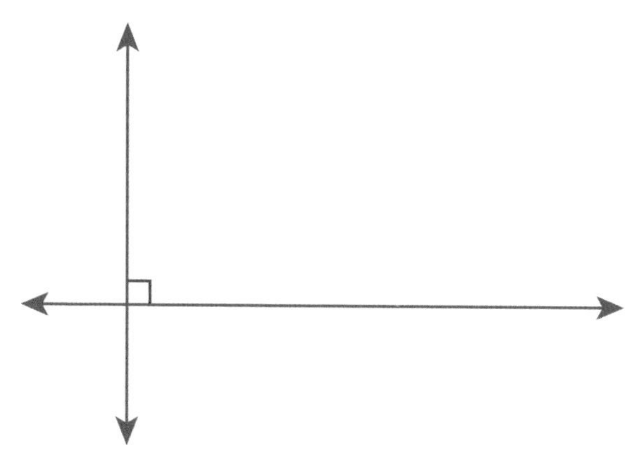

Skew lines are lines that do not lie in the same plane, and do not intersect.

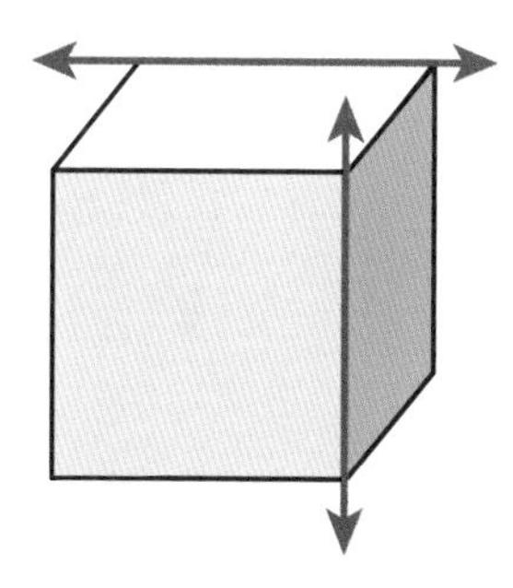

A **transversal** is a line that intersects two or more lines. In the figure below, line t is a transversal. (In everyday language, transverse means "lying across.")

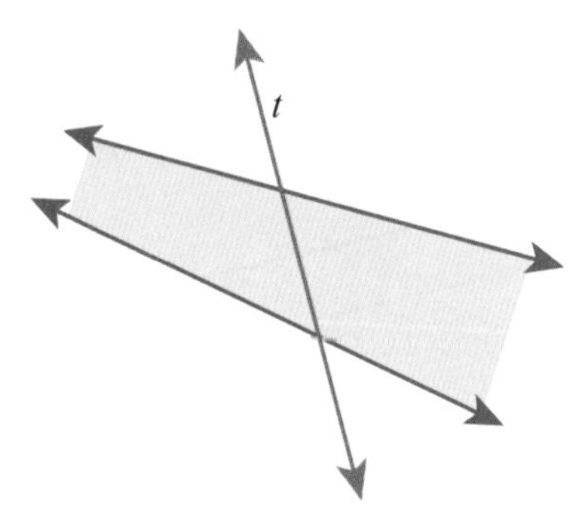

When two lines are crossed by a transversal, the **interior** is the region between two lines being discussed (other than the transversal), and the **exterior** is the region outside the interior. In the figure above, the interior is shaded.

There are several kinds of angle pairs that are formed by transversals, and it is useful to know their names:

Alternate interior angles lie in the interior region, and on opposite sides of the transversal.

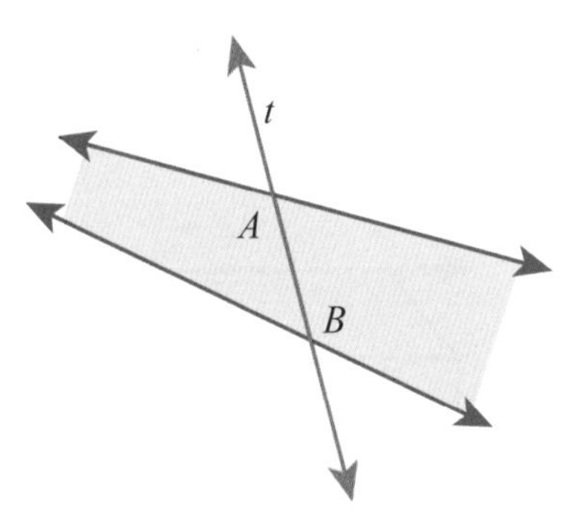

Alternate exterior angles lie in the exterior region, and on opposite sides of the transversal.

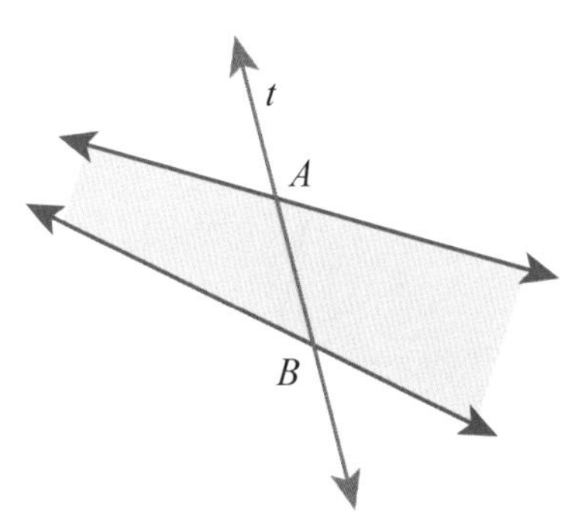

Same-side interior angles lie in the interior region, and on the same side of the transversal.

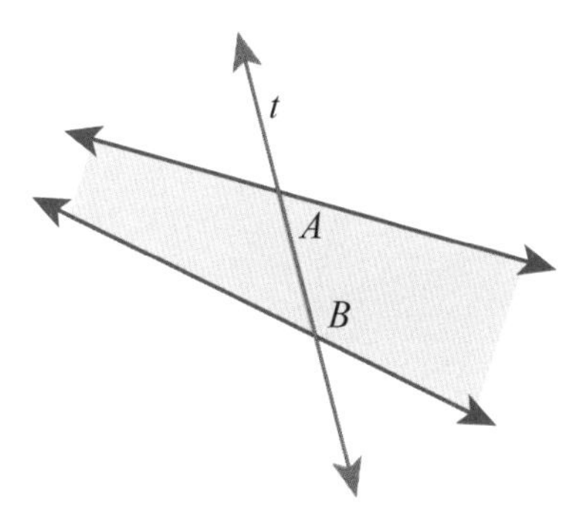

Same-side exterior angles lie in the exterior region, and on the same side of the transversal.

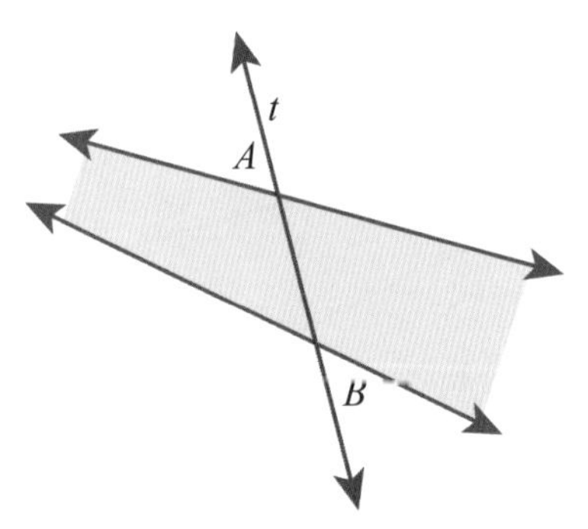

Corresponding angles are in corresponding positions, and on the same side of the transversal. For example, in the figure below, the corresponding angles shown are both in the "top right" corner relative to their intersection points.

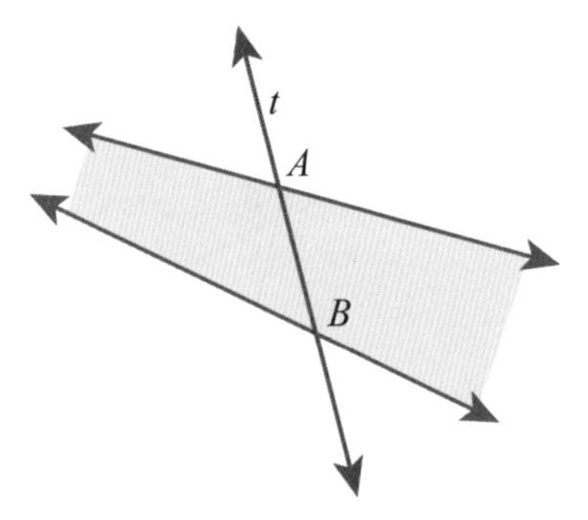

PROPERTIES OF PARALLEL LINES

When two lines are **parallel** and cut by a transversal, special angle pairs are formed. (To be clear, the properties below apply to *parallel lines only.*)

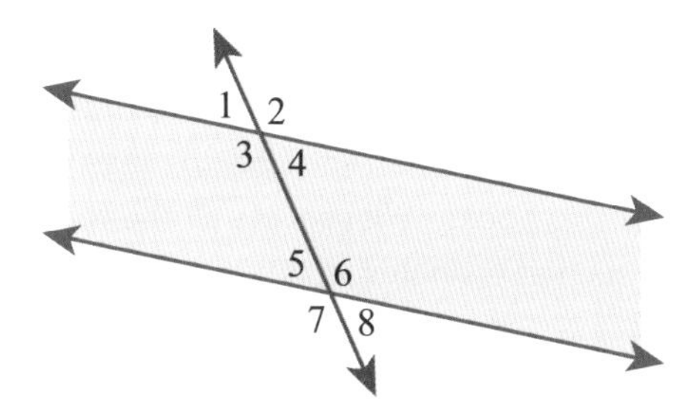

Rules	Examples
Alternate interior angles are *congruent.*	$\angle 3$ and $\angle 6$ are congruent.
	$\angle 4$ and $\angle 5$ are congruent.
Alternate exterior angles are *congruent.*	$\angle 1$ and $\angle 8$ are congruent.
	$\angle 2$ and $\angle 7$ are congruent.
Same-side interior angles are *supplementary.*	$\angle 3$ and $\angle 5$ are supplementary.
	$\angle 4$ and $\angle 6$ are supplementary.
Same-side exterior angles are *supplementary.*	$\angle 1$ and $\angle 7$ are supplementary.
	$\angle 2$ and $\angle 8$ are supplementary.
Corresponding angles are *congruent.*	$\angle 1$ and $\angle 5$ are congruent.
	$\angle 2$ and $\angle 6$ are congruent.
	$\angle 3$ and $\angle 7$ are congruent.
	$\angle 4$ and $\angle 8$ are congruent.

Fortunately, if you can remember just one of these rules accurately, you can figure out the others by using what you know about straight angles and/or vertical angles. In fact, if you know that just *one* angle in a diagram of parallel lines with a transversal, then you can figure out all eight angles! Note that the "alternate interior angles" rule is the one most commonly referenced, so that would be a good rule to memorize.

INTRO TO AREA AND PERIMETER

Area and perimeter are how we measure the size of figures. If you're taking a geometry class, you can expect to spend a lot of time on these concepts, so they are quite important. In this section, we'll provide an overview of what these terms mean and how to measure certain figures. In other chapters of this book, we'll provide all the formulas you'll need for specific shapes such as triangles, quadrilaterals, and circles.

PERIMETER

The **perimeter** of any shape is the sum of all of its side lengths. If the side lengths are given, then all you need to do is add them up!

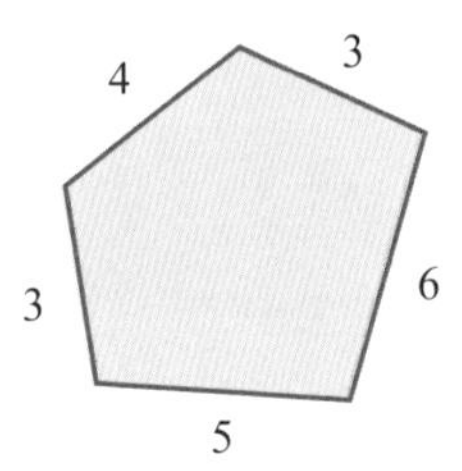

In the figure above, the perimeter is 4 + 3 + 6 + 5 + 3 = **21.**

To find perimeter from a grid, count the units along the entire border of the figure.

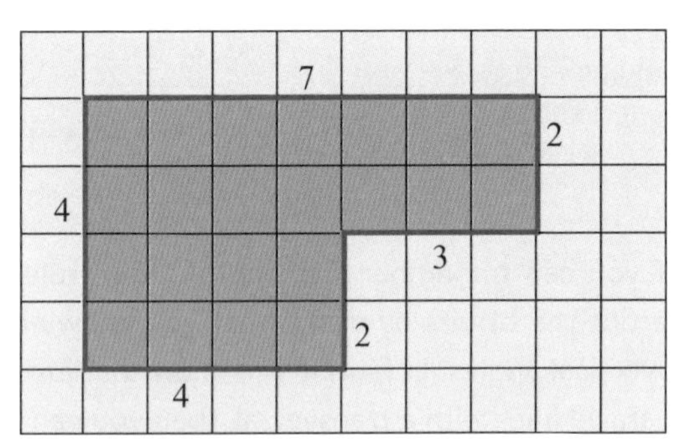

In the figure above, the perimeter is 7 + 2 + 3 + 2 + 4 + 4 = **22.**

Note that the perimeter includes only the outer edges. If a figure includes interior lines, they should *not* be included in the perimeter.

Perimeter is measured in units of *length*. These units include inches, feet, centimeters, meters, and so on. Generally, it's very important to include the units in your work (e.g., "2 feet"). If the type of unit is not specified, then we can use the term "units" as a sort of placeholder.

TRICKY PERIMETER PROBLEMS

Some problems appear complex or seem to have too many unknowns. To solve them, think about how you can figure out any unknown side lengths first.

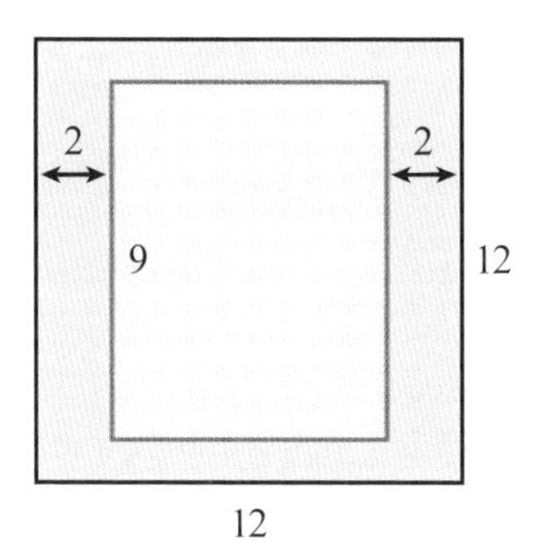

To find the perimeter of the inner white rectangle above, we'll need to find a few missing side lengths first. Take note of the given information: the height of the white rectangle is 9, but we do not yet know the width. The larger rectangle has height 12 and width 12. The shaded border has a width of 2 on the left and right sides. Therefore, the width of the white rectangle must be **8**, since 12 – 2 – 2 = 8. The perimeter, then, is 9 + 9 + 8 + 8 = **34**.

You will learn more about rectangles and other quadrilaterals in Chapter 3.

Note: if the length of a diagonal line is not given, do not assume you know its exact length. For example, the diagonal line below does *not* have a length of exactly 1 unit. If you are asked to estimate, just take your best guess, but don't forget to use a word like *approximately*.

You will learn about calculating diagonal line lengths in Chapter 2.

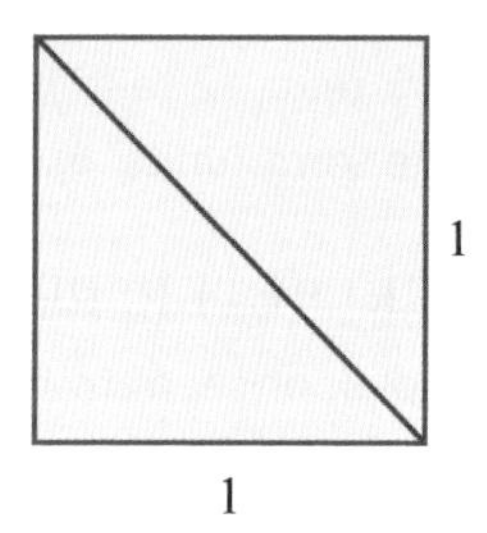

The diagonal is *longer* than 1 unit!

AREA

The **area** of a shape refers to the amount of space it takes up in the plane. Area is measured in *square units*, such as square inches (in^2), square centimeters (cm^2), and so on. A *square unit* is a square whose sides are 1 unit long. For example, a *square inch* is 1 inch $\times$ 1 inch, and a *square centimeter* is 1 cm $\times$ 1 cm.

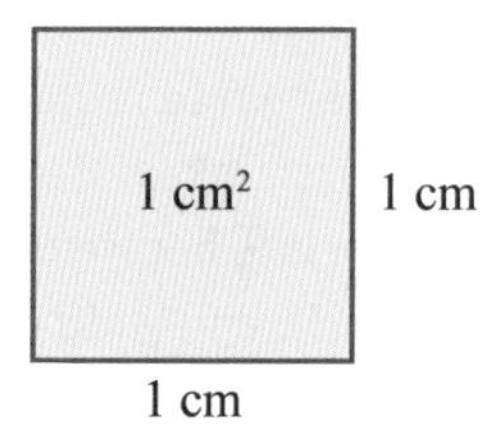

To find area of a figure on a grid, count all the complete squares inside the figure.

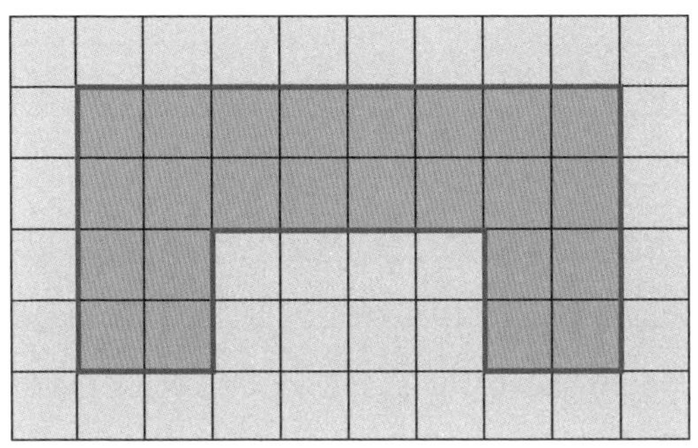

The area of the shaded figure above is **24**.

If a rectangle is divided diagonally at the corners, the diagonal divides the rectangle into two congruent triangles. Each triangle represents one-half of the rectangle. When you see a diagonal line in a diagram, first find the rectangle and calculate the area of that rectangle. To find the area of one of the triangles, simply divide the area of the rectangle by 2.

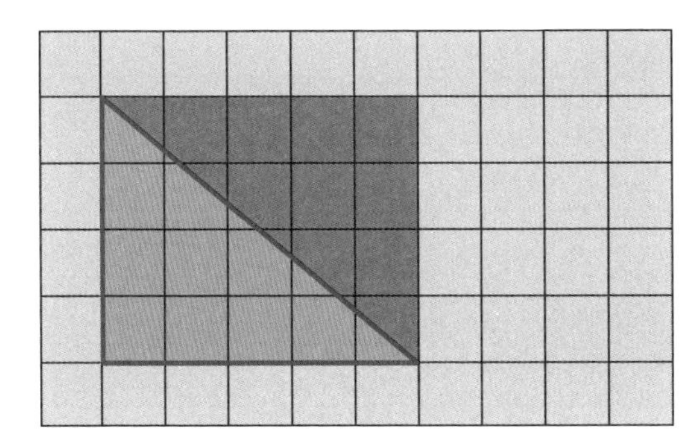

The blue triangle above is half of a rectangle, with height 4 and width 5. The area of the rectangle is 20, so the area of the triangle is **10**.

For other complicated shapes with partial squares, if you don't know a formula, then you might need to estimate. A good estimate is $\frac{1}{2}$ square unit for every partial unit inside the figure. When estimating, don't forget to use a word like *approximately*.

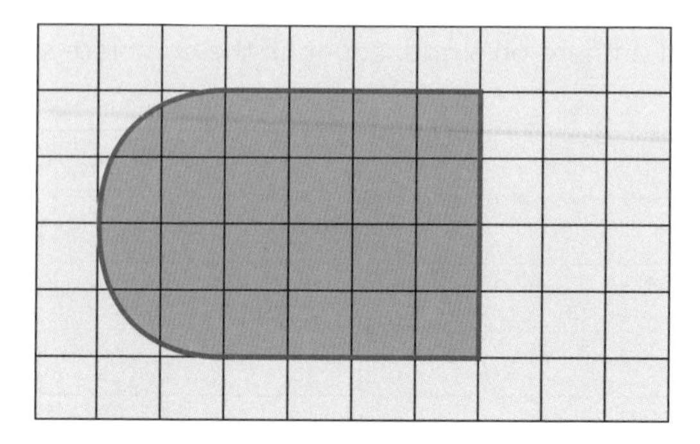

In the shaded figure above, the shape appears to be made from a rectangle and a half-circle combined. You can count 16 complete squares inside the rectangular portion, plus 2 complete squares inside the half-circle portion. Then, within the half-circle, there are 6 partial squares, which we can estimate as $\frac{1}{2}$ square unit each. This gives an estimated area of 3 for all the partial squares. The estimated total area of the shaded figure is 16 + 2 + 3 = **21**.

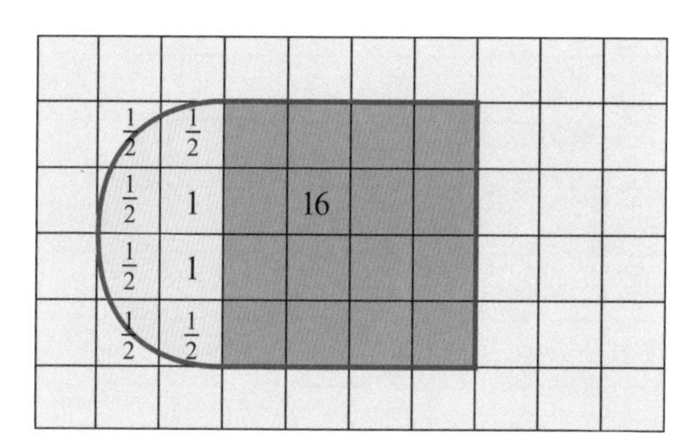

TRICKY AREA PROBLEMS

Some problems require you to find the area of a "complex figure." A complex figure is just a combination of different shapes. To find its area, you'll need to look for the simpler shapes that make the figure, and add the areas together.

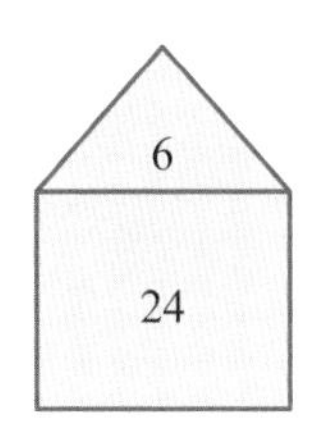

In the figure above, if we found the area of the rectangular portion to be 24, and the area of the triangular portion to be 6. Then the total area for the figure is 24 + 6 = **30**.

RELATIONSHIP BETWEEN AREA AND PERIMETER

When comparing two shapes, a larger area does *not* always mean a larger perimeter, and vice versa!

In this example, the rectangle on the right has the same width, but a greater length than the rectangle on the left. This does, in fact, result in a larger area and a larger perimeter.

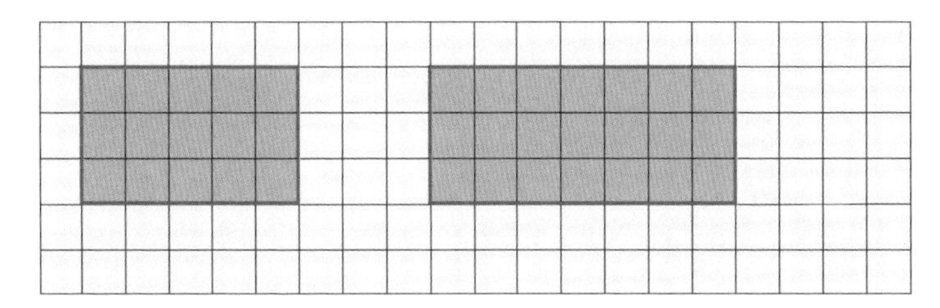

In this example, the rectangle on the right has a greater length but short width than the rectangle on the left. The perimeters of both rectangles are now the same (16 units), but the area of the rectangle on the right is now smaller than that of the rectangle on the left.

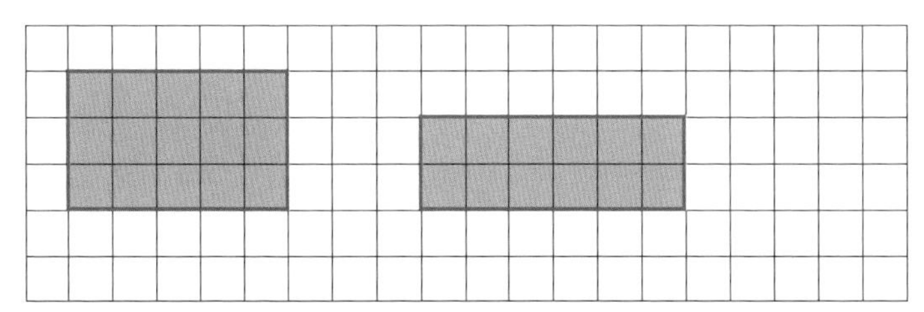

CHAPTER

PROPERTIES OF TRIANGLES

© istockphoto.com / Yevhen Roshchyn

Triangles are the building blocks of much of geometry. Many of the complex theorems of advanced geometry and trigonometry are rooted in the basic properties and definitions of triangles. Though triangles may appear to be among the simplest of geometric figures, that very simplicity creates a wide variety of properties and types to study.

CHAPTER CONTENTS

In this chapter, we'll discuss the properties of triangles. Learning about triangles will be important for learning the properties of other shapes. Furthermore, triangles are fundamental to the study of Trigonometry.

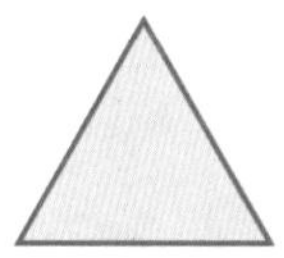

A triangle is a type of **polygon**. A polygon is a closed figure with three or more sides. The definition of a **triangle** is a polygon with exactly 3 sides.

You will learn more about polygons in the next chapter.

Each corner of a polygon is called a **vertex** (plural: **vertices**). A triangle has 3 vertices.

In any triangle, the sum of the three angle measures is **180°**.

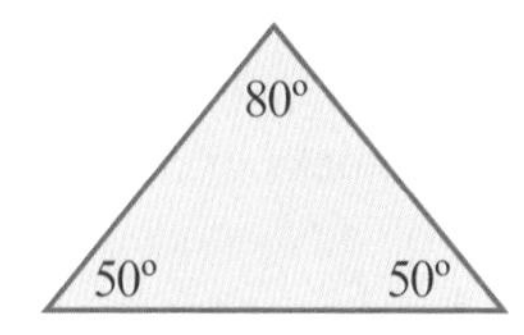

BONUS

If you make a triangle out of paper, you can cut off the three corners, and place them together. Since their sum is 180°, they will form a straight line!

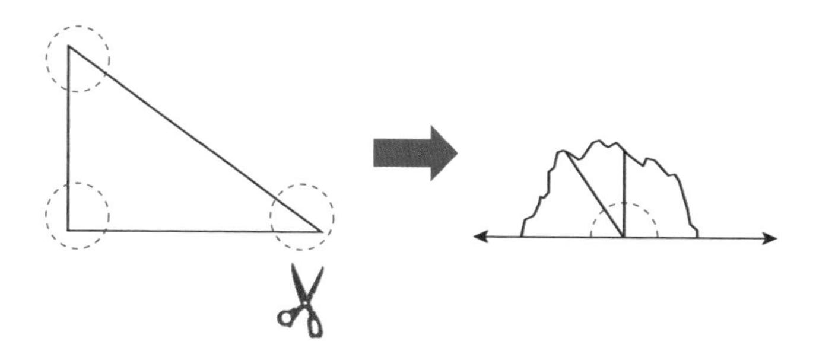

TYPES OF TRIANGLES

We can classify triangles by their angles.

An **acute** triangle has three acute angles.

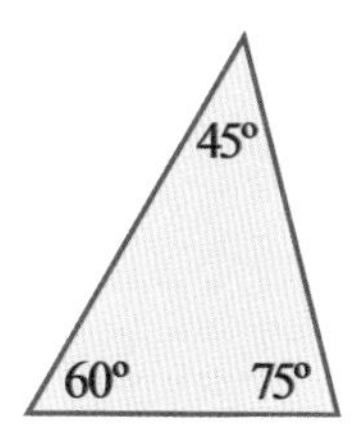

A **right** triangle has one right angle (and two acute angles).

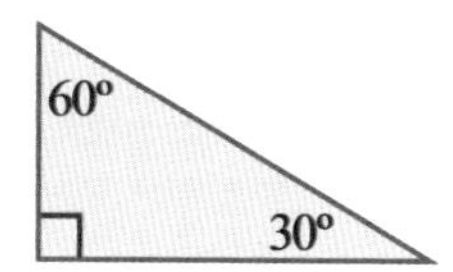

An **obtuse** triangle has one obtuse angle (and two acute angles).

Why is it impossible for a triangle to have two obtuse angles? Or two right angles?

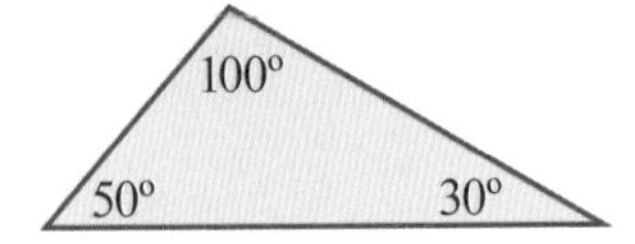

Another way to classify triangles is by their sides.

An **equilateral** triangle has three congruent sides. Also, all equilateral triangles have three congruent angles (each angle is 60°).

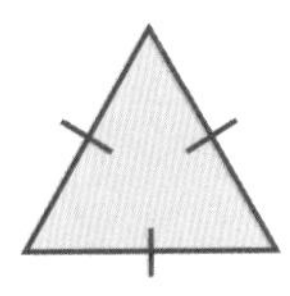

REMEMBER...

Congruent means exactly the same shape and size.

An **isosceles** triangle has two congruent sides. It also has two congruent angles.

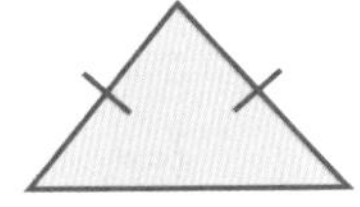

A **scalene** triangle has no congruent sides—all three sides are different. It has no congruent angles.

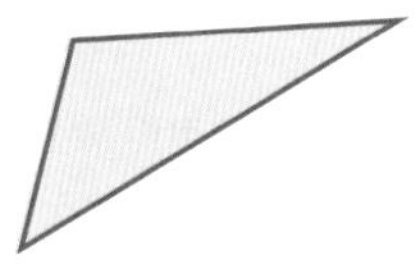

ASK YOURSELF...

Are all equilateral triangles acute?
Are all acute triangles equilateral?

PERIMETER AND AREA

The perimeter of a triangle is the sum of its three side lengths. If the three sides are known, simply add them!

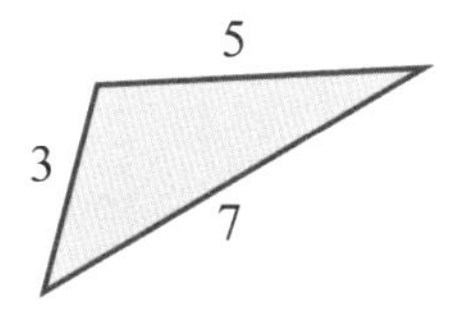

The perimeter of the triangle above is 3 + 5 + 7 = **15.**

To find the area of a triangle, use the following formula:

AREA OF A TRIANGLE

$$A = \frac{1}{2}bh$$

where b is the base, and h is the height of the triangle.

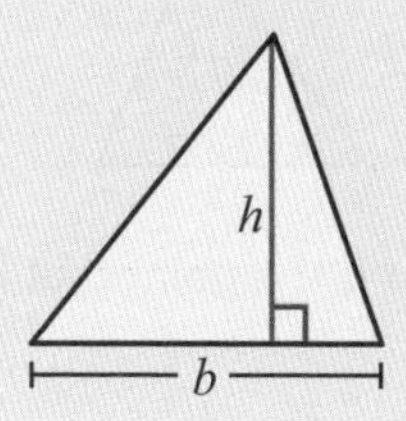

Note that the height of a triangle (also known as **altitude**) is *perpendicular* to the base. It is *not* necessarily the same length as any of the sides.

BISECTORS, ALTITUDES, MIDSEGMENTS, AND MEDIANS

A **bisector** is a line that divides a figure exactly in half. An **angle bisector** divides an angle in half, and a **segment bisector** divides a line segment in half.

A **perpendicular bisector** is a segment bisector, *and* it is perpendicular to the segment. Every triangle has three perpendicular bisectors—one for each side.

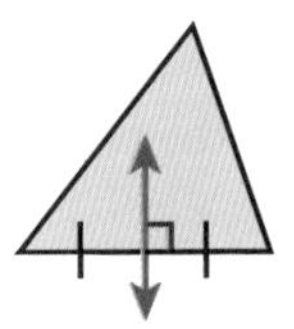

An **altitude** of a triangle is a segment that is perpendicular to one side of a triangle, and intersects the opposite vertex. Every triangle has three altitudes.

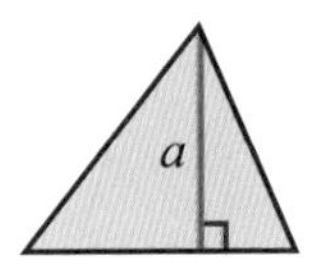

A **midpoint** of a line segment is the point that is exactly in the middle of the segment.

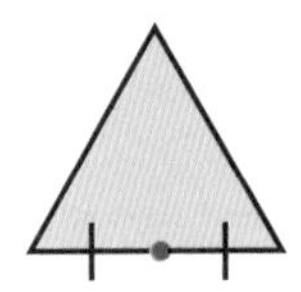

A **median** of a triangle is a segment that connects a midpoint of one side and its opposite vertex. Every triangle has three medians.

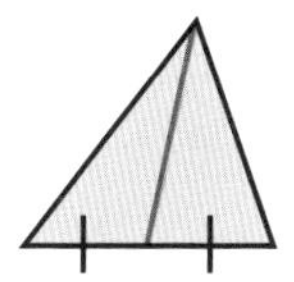

In a triangle, a **midsegment** is a segment connecting two midpoints of the triangle. Every triangle has three midsegments.

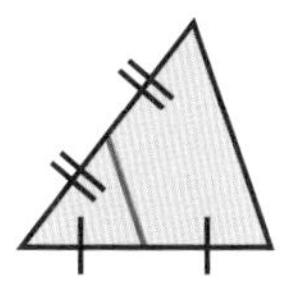

When a midsegment is formed, it is always parallel to the third side of the triangle, and it is half as long.

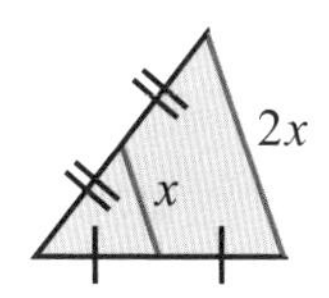

DIFFERENT "CENTERS" OF A TRIANGLE

Did you know that there are many ways to define the "center" of a triangle?

Previously, we reviewed the different segments that divide up a triangle—midsegments, angle bisectors, and perpendicular bisectors. These segments are interesting because of how they intersect. These intersection points are how we define the different centers of a triangle.

The **circumcenter** of a triangle is the intersection of its perpendicular bisectors.

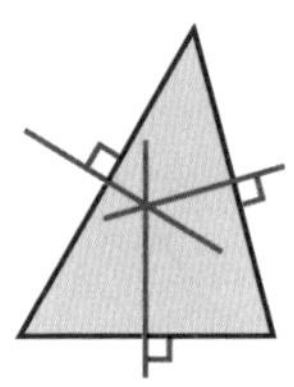

From the circumcenter, you can draw a circle that intersects all three vertices of the triangle. That circle is called the **circumcircle**, because it is *circumscribed* around the triangle.

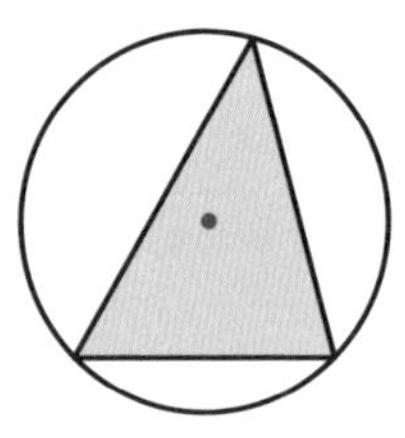

The circumcenter can be inside, outside, or on the triangle.

The **incenter** of a triangle is the intersection of its angle bisectors.

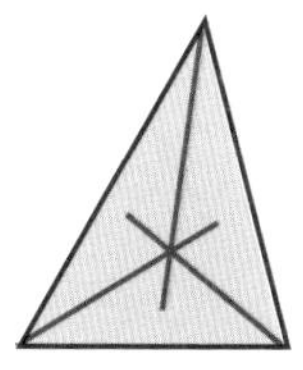

From the incenter, you can draw a circle that is tangent to all three sides of the triangle. That circle is called the **incircle**, because it is *inscribed* in the triangle.

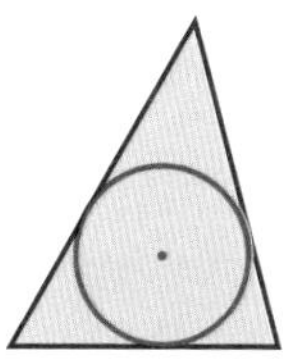

The incenter is always inside the triangle.

The **centroid** of a triangle is the intersection of its medians.

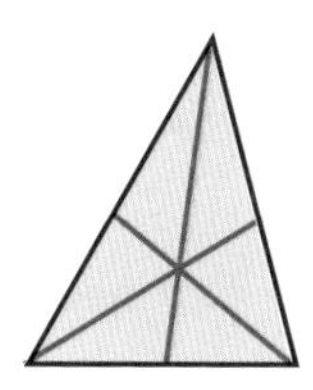

The centroid is also known as the *center of gravity* of a triangle, because it is the point where the "mass" of a triangle is balanced. That is, if you have a physical cutout of a triangle, you should be able to balance it on the centroid.

The centroid is always inside the triangle.

The **orthocenter** of a triangle is the intersection of its altitudes.

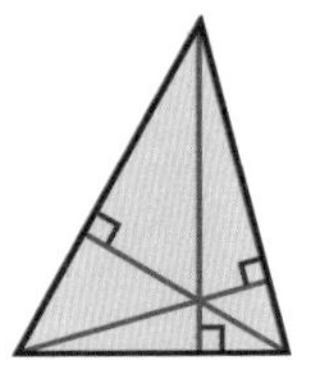

The orthocenter can be inside, outside, or on the triangle.

In Chapter 6 of this book, you will learn different techniques for *constructing* the centers of a triangle. To summarize, this is what you will need to do:

To find this...	Draw these...
Circumcenter	*Two* perpendicular bisectors
Incenter	*Two* angle bisectors
Centroid	*Two* medians
Orthocenter	*Two* altitudes

Once you correctly draw the two necessary segments, just look for their intersection point.

OTHER TRIANGLE PROPERTIES

THE PYTHAGOREAN THEOREM

The Pythagorean Theorem is used to solve for an unknown side length in a **right triangle**.

Note: in a right triangle, the longest side is called the **hypotenuse**, and it is always opposite the right angle. The two shorter sides are called **legs**.

PYTHAGOREAN THEOREM

In any right triangle,

$$a^2 + b^2 = c^2$$

where a and b are the legs, and c is the hypotenuse of the triangle.

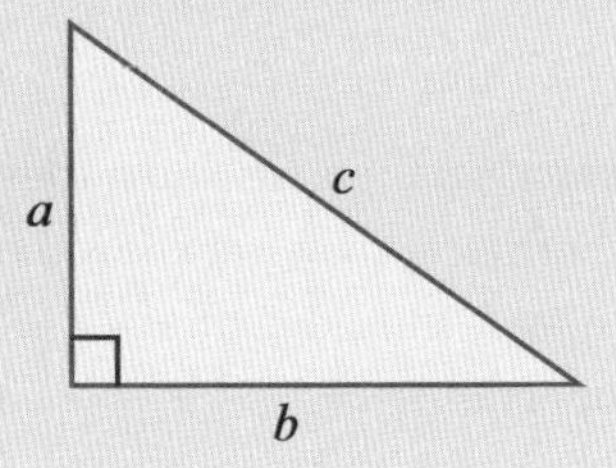

A **Pythagorean Triple** is a combination of three integers that satisfies the Pythagorean equation. Try memorizing the following Pythagorean triples to make right triangle problems easier. There are many more, but these are some of the most common ones (and the lowest values) that you'll encounter:

$$3^2 + 4^2 = 5^2$$

$$5^2 + 12^2 = 13^2$$

$$7^2 + 24^2 = 25^2$$

Furthermore, any *multiple* of a known Pythagorean Triple is also a Pythagorean Triple. For example, 6-8-10 is a multiple of 3-4-5, and 10-24-26 is a multiple of 5-12-13.

SPECIAL RIGHT TRIANGLES

There are two special cases for right triangles that are useful to memorize.

45°-45°-90° Triangle

If a right triangle is *isosceles*, then we know that it has two congruent angles, both 45°. Furthermore, since we know the angles, then we know the triangle's shape. In fact, the sides of a 45°-45°-90° triangle always have the following proportions:

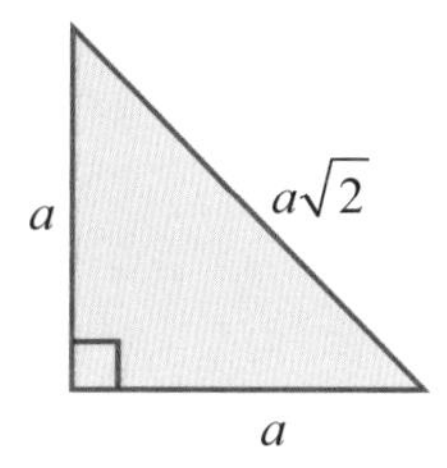

One way to prove this is with the Pythagorean Theorem. Since the triangle is isosceles, and has two congruent sides, we can set up the Pythagorean equation as follows:

$a^2 + a^2 = c^2$	*a* is used twice, since the two side lengths are equal
$2a^2 = c^2$	simplify the left side
$\sqrt{2a^2} = \sqrt{c^2}$	take the square root of both sides
$a\sqrt{2} = c$	simplify

Therefore, the hypotenuse of an isosceles right triangle has a length equal to the leg multiplied by $\sqrt{2}$. You can use this formula to simplify problems when you know that the right triangle is isosceles.

Note that this type of triangle is also formed by dividing a square in half diagonally. Therefore, this formula is useful when you need to know the diagonal of a square.

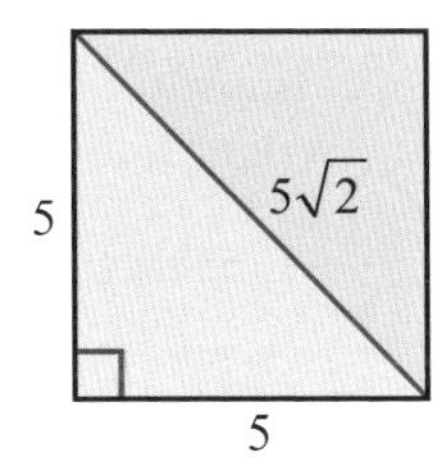

30°-60°-90° Triangle

If a triangle has the angles 30°-60°-90°, then its sides have the following proportions:

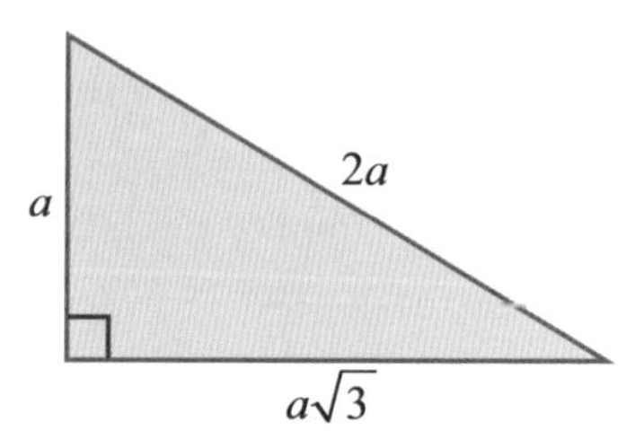

This type of triangle is formed by dividing an equilateral triangle in half by its altitude. Therefore, we know that the shortest leg is half the length of the hypotenuse.

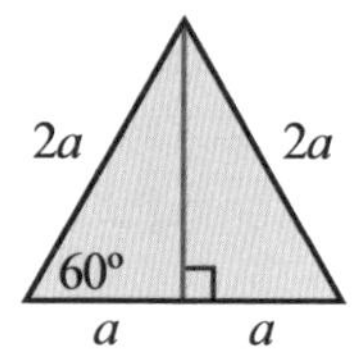

If we apply the Pythagorean Theorem, we find the length of the third side:

$a^2 + b^2 = (2a)^2$	substitute $(2a)$ for c on the right side
$a^2 + b^2 = 4a^2$	simplify the right side
$b^2 = 3a^2$	subtract a^2 from both sides
$b = \sqrt{3}a$	simplify

This formula is useful when you need to know the height of an equilateral triangle, or with other figures containing 30° or 60° angles.

LONGEST SIDE AND LARGEST ANGLE

The largest side of a triangle is opposite its largest angle. (Also, the largest angle is opposite the largest side.)

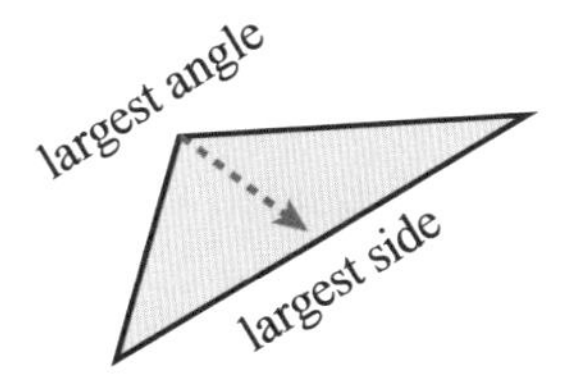

Likewise, the smallest side of a triangle is opposite its smallest angle.

TRIANGLE INEQUALITY THEOREM

The Triangle Inequality Theorem is also known as the **Third Side Rule**. It says that in a triangle, the sum of two side lengths is always greater than the length of the third side.

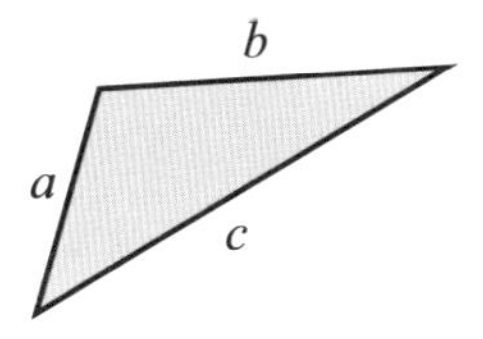

For the figure above, each of the following statements must be true:

- $a + b > c$
- $a + c > b$
- $b + c > a$

Similarly, the difference of two side lengths is always less than the length of the third side.

- $b - a < c$
- $c - a < b$
- $c - b < a$

In other words, find the sum of the two known sides, and also find the difference of the two known sides. The length of the third side has to be between these two values.

The Triangle Inequality Theorem is useful when you know two sides but have no other information. You can use this rule to find a *range* of possible side lengths for the third side.

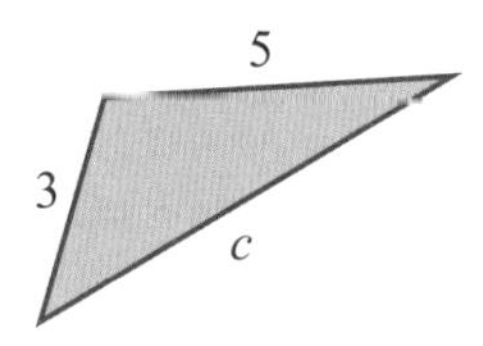

For the figure above, we know that $c < 8$, because $c < 3 + 5$.

Additionally, we know that $c > 2$, because $5 - 3 < c$.

Therefore, $2 < c < 8$.

CHAPTER

PROPERTIES OF QUADRILATERALS AND OTHER POLYGONS

Increasing the number of sides in a polygon does not always increase the complexity. The properties of these figures will follow directly from the properties of triangles.

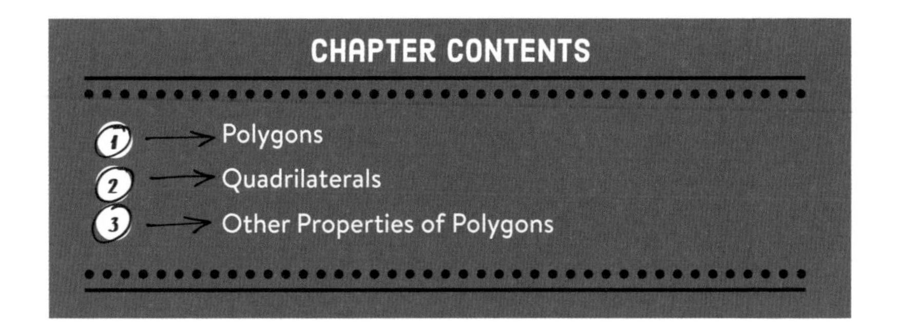

In this chapter, we'll discuss the basic properties of different shapes.

POLYGONS

Once more, a **polygon** is a closed figure with three or more sides. Each corner of a polygon is called a **vertex** (plural: **vertices**).

Note: each side of a polygon intersects exactly two other sides (at the vertices). That is, the sides do not cross each other in the middle.

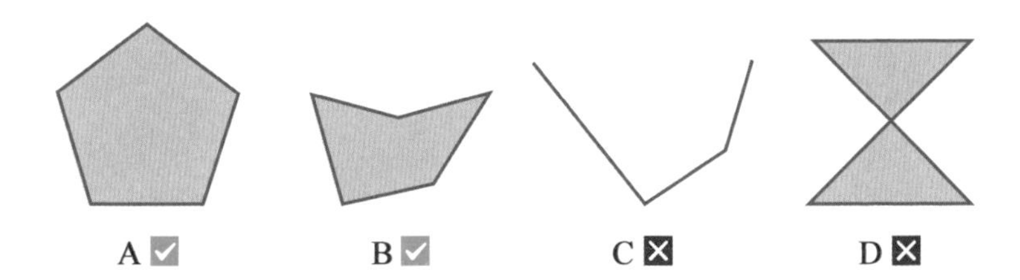

Figures A and B are polygons, while Figures C and D are not. Notice that a polygon's vertices can point "inward" or "outward."

Figure C is not a polygon because it is not closed. Figure D is not a polygon because the sides intersect each other. Note that this type of shape is sometimes called a "complex polygon," which means that it is a combination of two or more polygons.

Some Names of Polygons

Number of Sides	Name
3	Triangle
4	Quadrilateral
5	Pentagon
6	Hexagon
7	Heptagon
8	Octagon
9	Nonagon
10	Decagon

A **diagonal** is a segment connecting two nonadjacent vertices of a polygon.

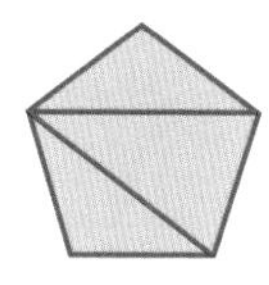

Note: a triangle does not have any diagonals.

A polygon is **convex** if no diagonals lie outside the polygon. It is **concave** if at least one diagonal lies outside the polygon.

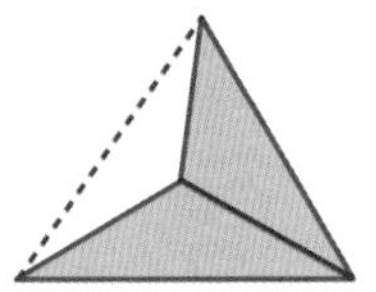

Note: all triangles are convex.

In everyday language, *convex* means "curved outward," and *concave* means "curved inward."

An **equilateral** polygon is one whose sides are all congruent.

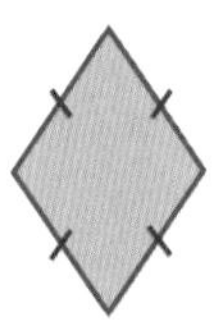

An **equiangular** polygon is one whose angles are all congruent.

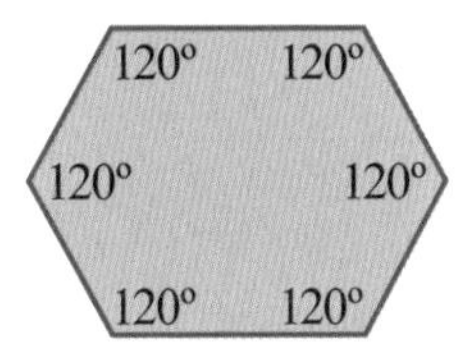

A **regular** polygon is both equilateral and equiangular.

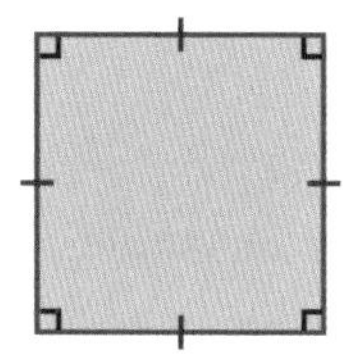

Note: triangles are a special case—if a triangle is equilateral, then it is equiangular, and vice versa. This is not always true for other polygons.

There is a formula to find the sum of the measures of the angles of any polygon. This is useful for finding an unknown angle measure.

Polygon Angle-Sum Theorem

For a polygon having n sides, the sum of its interior angle measures is $\mathbf{(n - 2)180}$.

That is, if a polygon has 5 sides, then its angle measures add up to

$$(5 - 2) \times 180$$
$$= 3 \times 180$$
$$= \mathbf{540}$$

Interestingly, the sum of the exterior angle measures is always the same for any polygon.

Polygon Exterior Angle-Sum Theorem

The sum of the measures of the exterior angles of a polygon is **360°**.

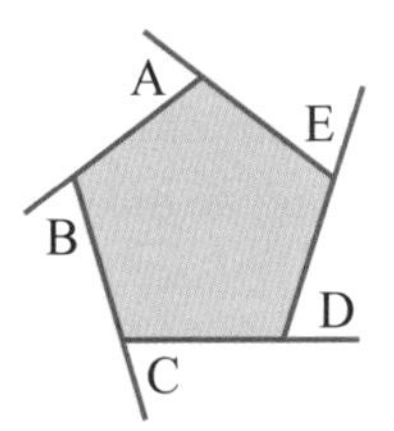

The **exterior angles** are formed by extending each side, so that it makes a linear pair with the interior angle there. Recall that a linear pair has a sum of 180°.

In this example, we start with a regular pentagon, where each interior angle measures **108°**. Therefore, each exterior angle is **72°** (108° + 72° = 180°).

The sum of these exterior angles is 72° × 5 = **360°**. This sum is the same no matter how many sides the polygon has.

ASK YOURSELF...

How are the Angle-Sum Theorem and the Exterior Angle-Sum Theorem related? In other words, does one theorem help explain the other?

QUADRILATERALS

A **quadrilateral** is any polygon with exactly 4 sides.

When discussing quadrilaterals, it is sometimes useful to refer to their **opposite** sides or angles. In a quadrilateral, opposite sides are not touching—that is, they do not share a vertex. Opposite angles do not share a side.

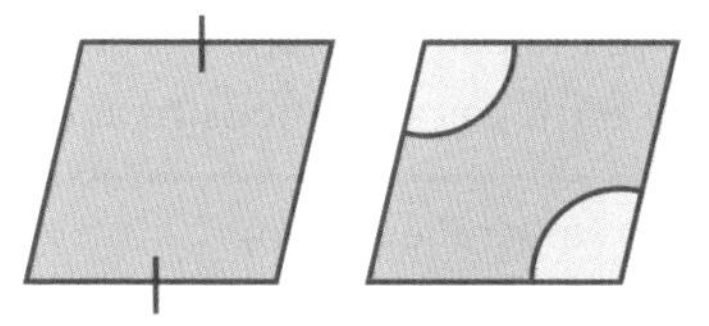

Consecutive sides are two sides sharing a vertex. Consecutive angles are two angles sharing a side.

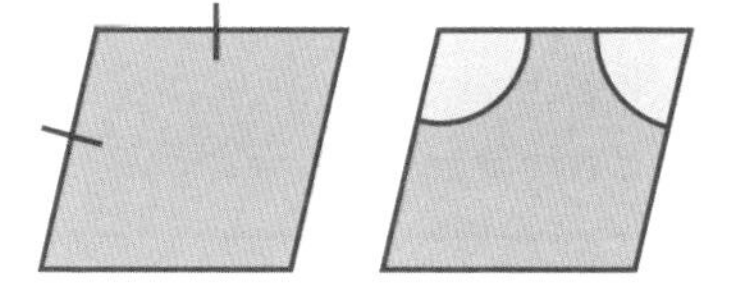

There are several special names for quadrilaterals having certain properties. Let's review:

Quadrilateral (any quadrilateral)	• is a polygon • has exactly 4 sides • has exactly 4 angles • sum of the 4 angle measures is 360°
Parallelogram	• is a quadrilateral • both pairs of opposite sides are parallel • both pairs of opposite sides are congruent • both pairs of opposite angles are congruent
Rhombus	• is a parallelogram • all four sides are congruent
Rectangle	• is a parallelogram • all four angles are right angles (90°)
Square	• is a parallelogram • is a rhombus • is a rectangle
Trapezoid	• is a quadrilateral • is not a parallelogram • has exactly one pair of parallel sides
Kite	• is a quadrilateral • is not a parallelogram • two pairs of consecutive sides are congruent • opposite sides are not congruent

Note that the *parallelograms* are **outlined in red**!

MORE PROPERTIES OF PARALLELOGRAMS

To find the *perimeter* of any parallelogram, add the four side lengths. In fact, since parallelograms always have two pairs of congruent sides, you can use the following formula:

Perimeter of a Parallelogram

$$P = 2l + 2w$$

$$P = 2(l + w)$$

where l and w are the length and width of the parallelogram.

To find the *area* of any parallelogram, use this formula:

Area of a Parallelogram

$$A = bh$$

where b is the base, and h is the height of the parallelogram

Note that the height of a parallelogram (also known as **altitude**) is *perpendicular* to the base. It is *not* necessarily the same length as any of the sides.

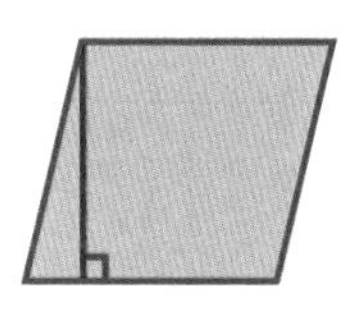

ASK YOURSELF...

How are the properties of a parallelogram related to the properties of parallel lines with a transversal?

In a parallelogram, each pair of consecutive angles is supplementary.

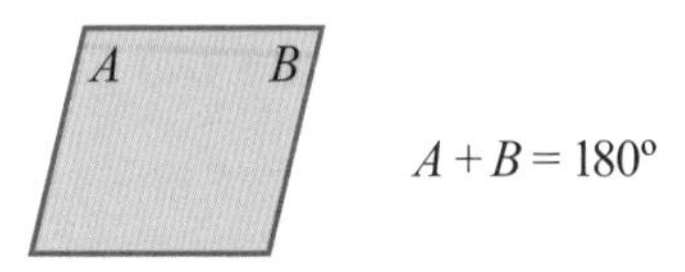

The diagonals of a parallelogram **bisect** each other.

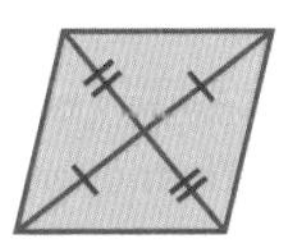

To **bisect** a shape means to divide into two congruent parts—in other words, to divide the figure exactly in half.

Bisecting a line segment divides it into two congruent segments.

Bisecting an angle divides it into two congruent angles.

Here are some additional properties to know:

For any **rhombus**, the diagonals are perpendicular.

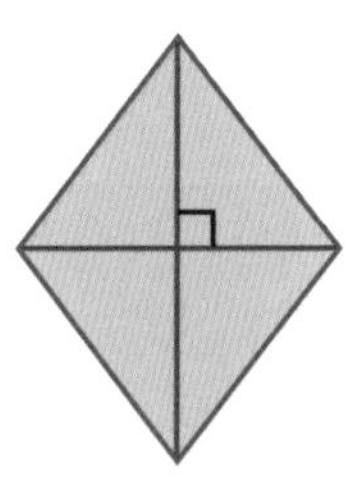

For any **rhombus**, the diagonals bisect the respective angles.

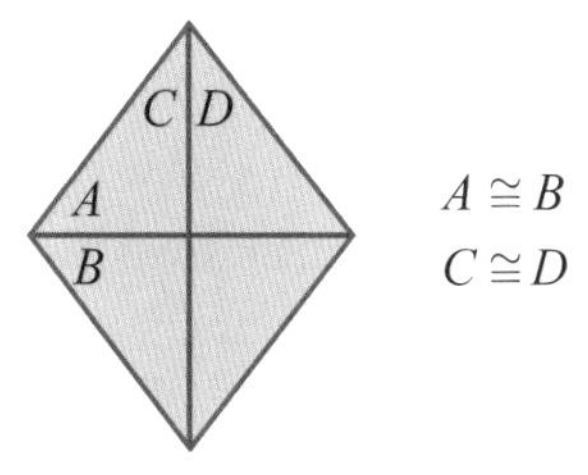

For any **rectangle**, the diagonals are congruent.

Consider the properties of parallelograms that we have learned. If we know that a figure is a parallelogram, then we can prove any of the properties previously discussed. On the other hand, what if we need to prove that a figure is a parallelogram? It's important to know that each of these properties also has a **converse**—a reverse property that is also true.

For any quadrilateral:

- If both pairs of opposite sides are congruent, then the quadrilateral is a parallelogram.
- If both pairs of opposite angles are congruent, then the quadrilateral is a parallelogram.
- If each pair of consecutive angles is supplementary, then the quadrilateral is a parallelogram.
- If the diagonals bisect each other, then the quadrilateral is a parallelogram.

That means that each of these properties alone is sufficient to *prove* that a quadrilateral is a parallelogram.

Furthermore, for any parallelogram:

* If the diagonals are perpendicular, then the parallelogram is a rhombus.
* If the diagonals bisect the angles, then the parallelogram is a rhombus.
* If the diagonals are congruent, then the parallelogram is a rectangle.

MORE PROPERTIES OF KITES AND TRAPEZOIDS

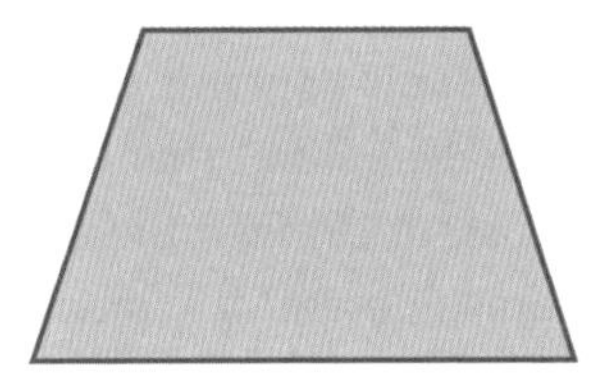

For a trapezoid, the parallel sides are called the **bases**, and the nonparallel sides are called the **legs**.

There are two pairs of **base angles**—angles which share a base.

An **isosceles trapezoid** is a trapezoid whose legs are congruent. In an isosceles trapezoid, each pair of base angles is congruent. Also, its diagonals are congruent.

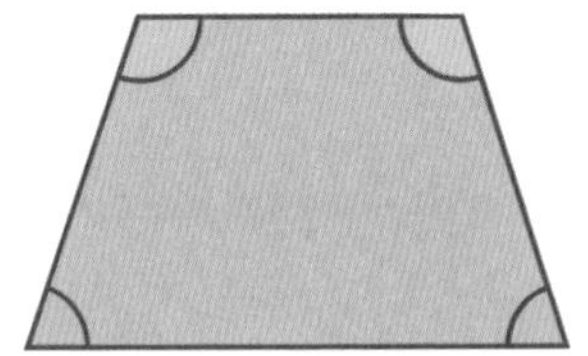

Area of a Trapezoid

$$A = (b_1 + b_2)h$$

where h is the height, and b_1 and b_2 are the two parallel bases of the trapezoid.

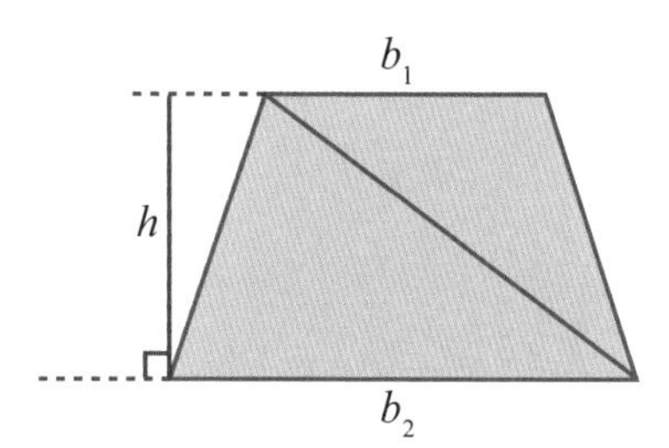

You can think of a trapezoid as two triangles formed from the trapezoid's diagonal. The two triangles have the same height, h, but different bases, b_1 and b_2. Therefore, their areas would be $\frac{1}{2}hb_1$ and $\frac{1}{2}hb_2$, respectively.

If you add the two together, you get $A = \frac{1}{2}hb_1 + \frac{1}{2}hb_2$. Using the distributive property, we can simplify to the formula above: $A = \frac{1}{2}h(b_1 + b_2)$.

If you ever forget the formula for a trapezoid, try calculating it as the sum of two triangles!

For any kite, the diagonals are perpendicular.

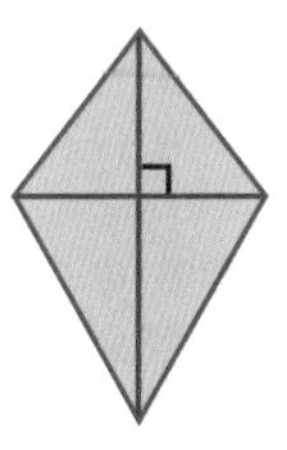

To find the area of a kite, you can use its diagonals. The same formula works for a rhombus as well (this is useful when you know the diagonals, but not the base and height).

Area of a Kite or Rhombus

$$A = \frac{1}{2}d_1d_2$$

where d_1 and d_2 are the lengths of the diagonals.

OTHER PROPERTIES OF POLYGONS

Recall that the **perimeter** of any polygon is the sum of its side lengths.

The **radius** of a regular polygon is the distance from the center to one vertex. If you draw a circumcircle around the polygon, then this is also the radius of the circle.

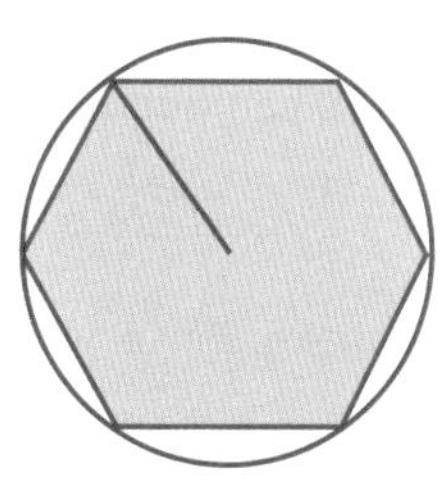

The **apothem** of a regular polygon is the perpendicular distance from the center to one side.

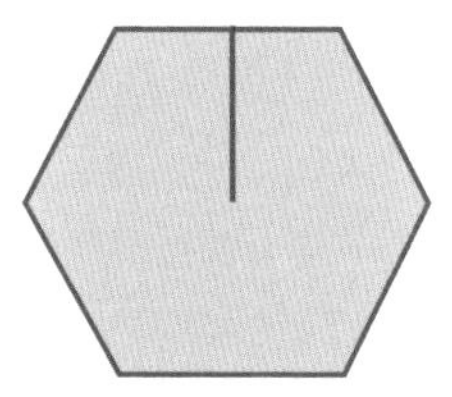

Area of a Regular Polygon

$$A = \frac{1}{2}aP$$

where a is the apothem, and P is the perimeter of the polygon.

You can think of a regular polygon as several triangles formed from the polygon's center.

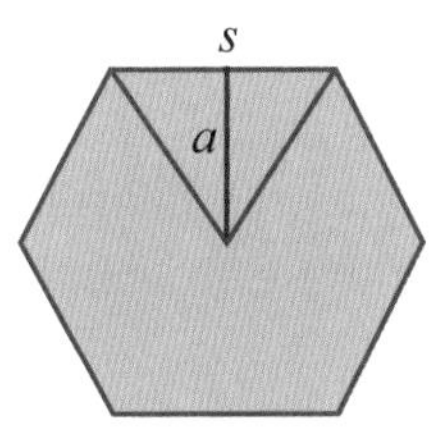

Therefore, to find the area, you can find the sum of the triangles' areas. Each triangle would have $A = \frac{1}{2}as$, where a is the apothem, and s is the side length. Now, it might make sense why the area of the polygon would be $A = \frac{1}{2}aP$.

CHAPTER

PROPERTIES OF CIRCLES

Circles don't have much variety: only the size can really change. However, the way that circles can be used can vary widely. Understanding the different parts of applications of circles is very important for a stronger understanding of geometry.

CHAPTER CONTENTS

A **circle** is a very recognizable shape, but its definition might be unfamiliar. In fact, a circle is defined as *the set of points located at a fixed distance from a given central point*. This basically just means that all the points on a circle are the same distance from the center.

The **radius** (plural: **radii**) of a circle is the distance from the center of the circle to a point on its edge.

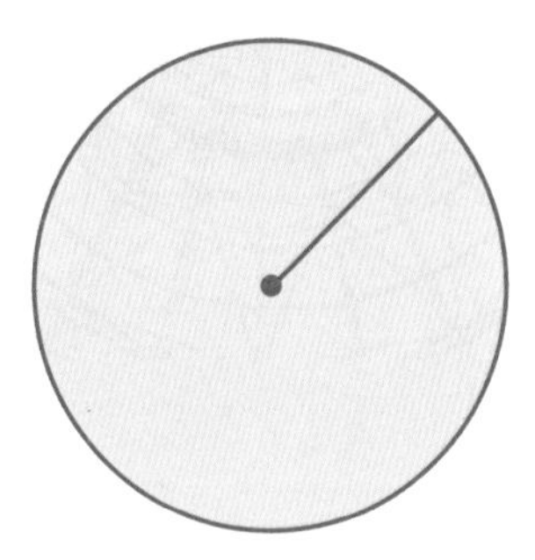

The **diameter** is equal to twice the radius. It is the distance between two opposite edges of the circle (and through the center).

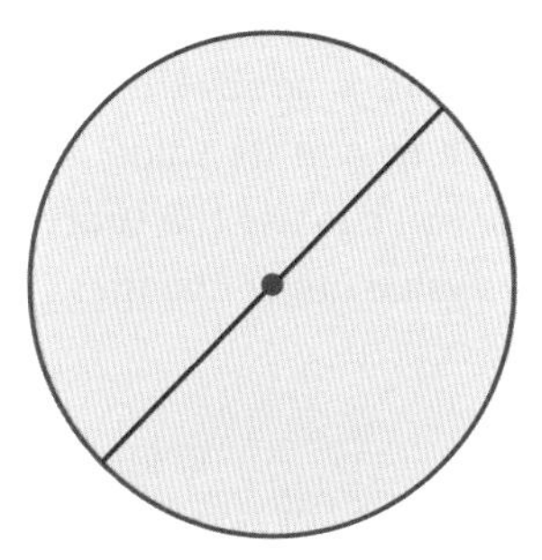

If two circles are congruent, then their radii and diameters are congruent.

Concentric circles are circles that share the same central point.

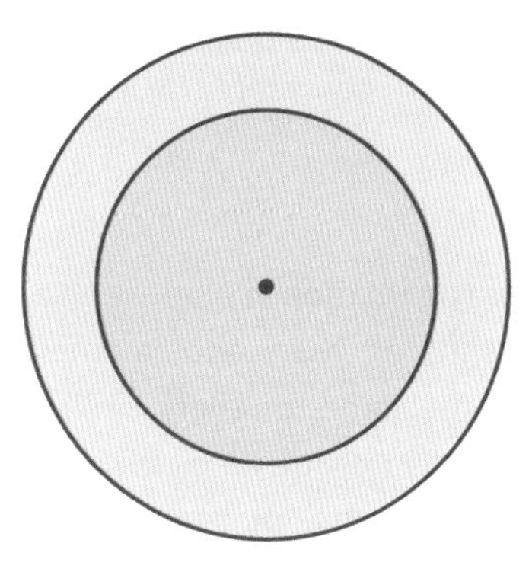

The distance between two concentric circles is the same all the way around. The “ring” or “donut” shape formed by two concentric circles is called an **annulus**.

AREA AND CIRCUMFERENCE

Pi (pronounced like "pie") is a special number in mathematics. Its symbol is the Greek letter **π**, and its value is approximately **3.14159**. (When doing calculations, you can use the shorter estimate **3.14**.) Pi is a *constant*, meaning that its value never changes.

The *exact* value of pi is an *irrational number*, meaning that its decimal value has infinitely many digits without any repeating pattern. When you are asked to give a decimal answer involving pi, you are actually giving an *estimate* by rounding.

If you are asked to give an *exact* answer involving pi, you would use the symbol π instead of its decimal value. For example, the value of (3 × π) can simply be written as 3π.

Pi is actually defined as the relationship between a circle's diameter and circumference. Thousands of years ago, mathematicians figured out that this relationship is the same for every circle! No matter how large or small a circle is, when you divide its circumference by the diameter, you will always get the same value, **pi**.

Pi

$$\pi = \frac{circumference}{diameter}$$

The value of π is approximately **3.14159.**

This gives us the formula for a circle's circumference, as shown below.

CIRCUMFERENCE

A circle's **circumference** is analogous to the *perimeter* of a polygon. That is, it is the distance all around the circle's edge. To find the circumference of a circle, use this formula:

Circumference of a Circle

$$C = 2\pi r$$

Or

$$C = \pi d$$

where r is the radius of a circle, and d is the diameter.

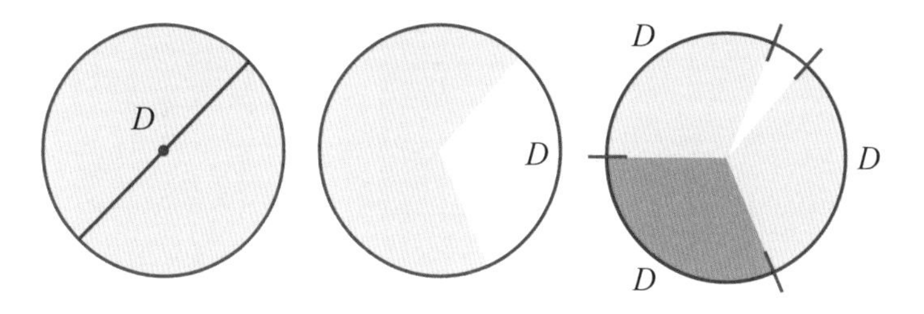

In other words, the length of the diameter can be found approximately 3.14 times around the circle's edge.

AREA

The area of a circle is given by the following formula:

Area of a Circle

$$A = \pi r^2$$

where r is the radius of a circle.

You can think of r^2 as a square with side length r. Imagine 3 such squares arranged over the circle. Since 3 is close to the value of π, the area of the three squares is close to the area of the circle.

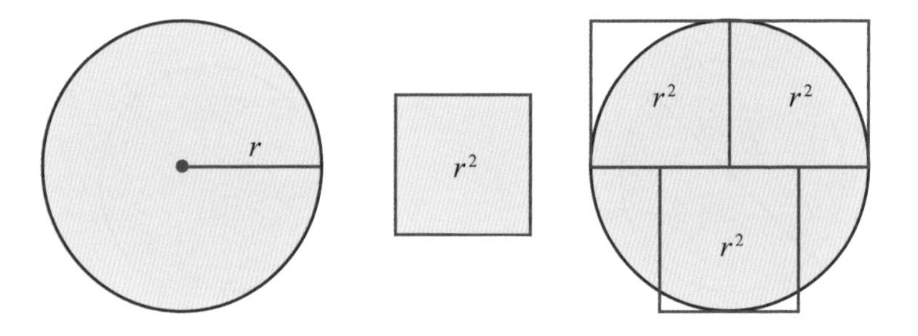

CARD

CARD is a mnemonic to remind you that all four of these elements of a circle—**C**ircumference, **A**rea, **R**adius, and **D**iameter—are related to one another. If you have just one of these values, you can easily solve for the other three.

A good strategy for any circle problem is to write C A R D on the paper, include any known values, and solve for the others.

Example: *The radius of a circle is 3.*

C

A

R = 3

D

From here, we know that the diameter is 6, because $D = 2r$.

The circumference is 6π, because $C = 2\pi r$.

And the area is 9π, because $A = \pi r^2$.

$C = 6\pi$

$A = 9\pi$

$R = 3$

$D = 6$

Usually, if the radius is unknown, it makes sense to solve for that first. The table below shows how to use the formulas to solve for radius.

FORMULA	TO SOLVE FOR RADIUS, DO...
$C = 2\pi r$	$C = (2\pi) \times r$ $\frac{C}{2\pi} = r$
$A = \pi r^2$	$A = \pi \times r^2$ $\frac{A}{\pi} = r^2$ $\sqrt{\frac{A}{\pi}} = r$
$D = 2r$	$D = 2 \times r$ $\frac{D}{2} = r$

CHORDS, SECANTS, AND TANGENTS

A **chord** is a line segment that connects any two points on a circle. The longest length for a chord is the diameter.

A **secant** also passes through two points on a circle, but it's a *line*—that is, it has no endpoints.

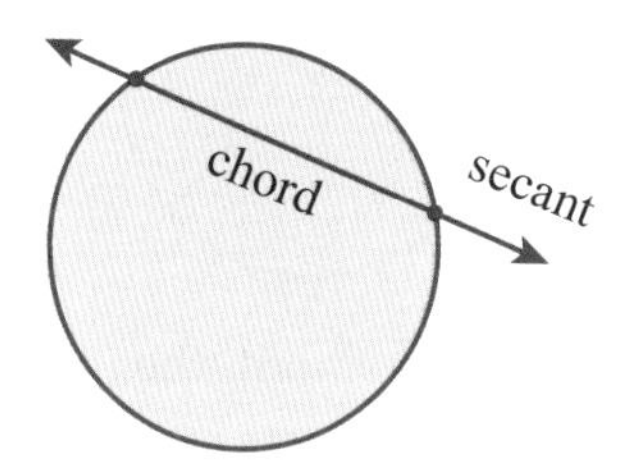

A **tangent** line (or segment) is one that touches the circle at exactly one point.

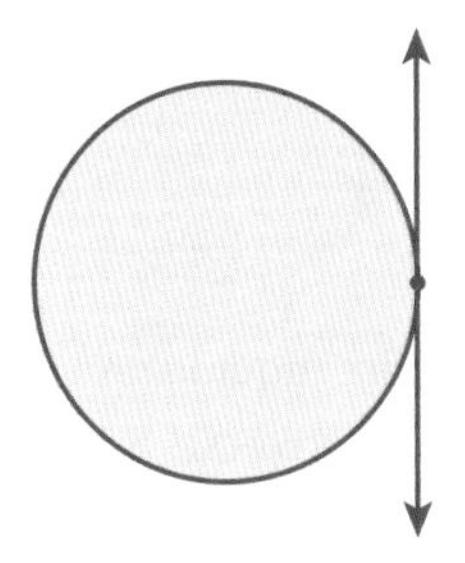

Here are some properties to know:

If a triangle is formed from a chord and the center of a circle, as shown, then the triangle is certainly isosceles. This is because *all radii in a circle are congruent*. In the figure, two of the triangle's sides are radii of the circle; therefore, they are congruent.

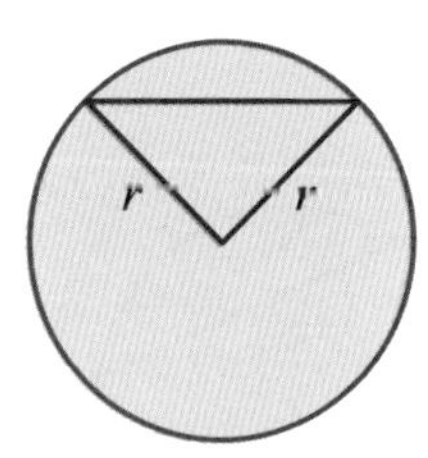

If a radius is perpendicular to a chord, then it bisects the chord. The converse is also true—if a radius bisects a chord, then it is perpendicular to the chord.

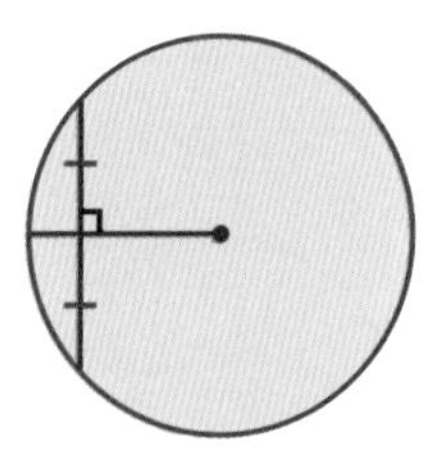

In a circle, if two chords are equidistant from the center, then the chords are congruent. The converse is also true—if two chords in a circle are congruent, then they are equidistant from the center.

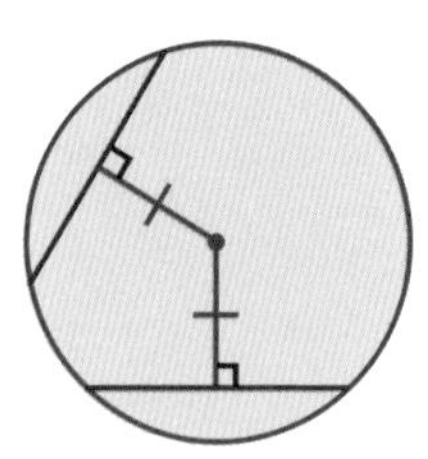

The perpendicular bisector of a chord always passes through the center of the circle.

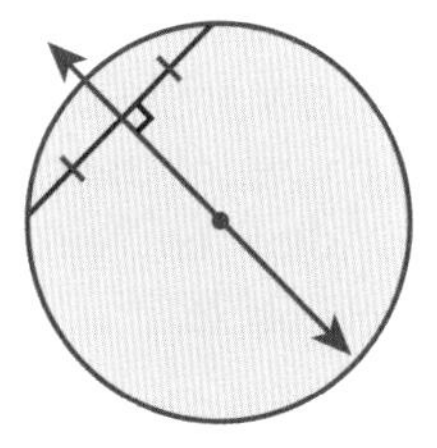

A line tangent to a circle intersects the circle at exactly one point. The intersection point is known as the **point of tangency**.

The radius intersects the point of tangency at a right angle (90°).

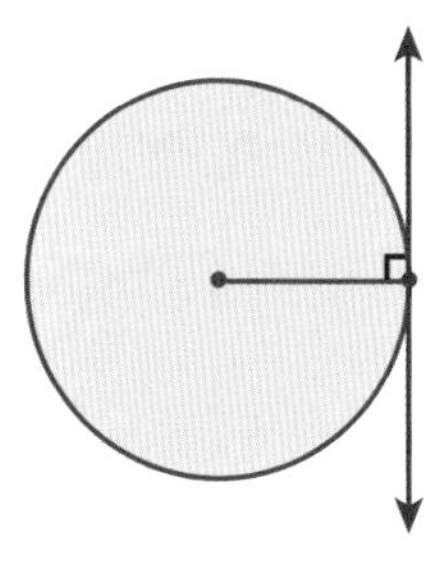

If two tangent lines intersect, then two congruent segments are formed—the distance from the intersection to the tangent point is the same for both segments.

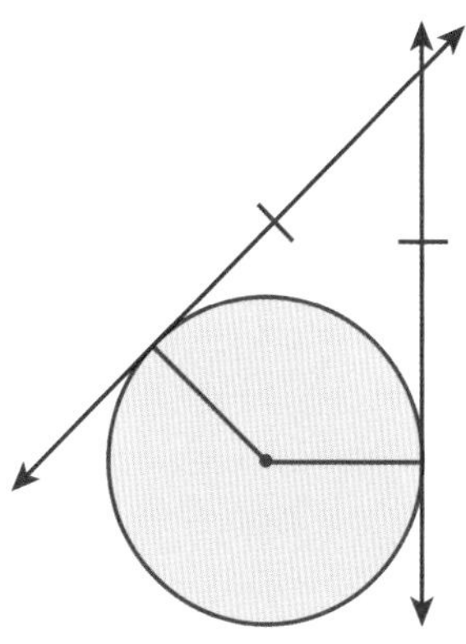

ARCS AND SECTORS

In a circle, a **central angle** is any angle that is drawn from the center (that is, the vertex is at the center of the circle). A circle has a total of 360°. That means that if you "slice" up a circle into many central angles, the sum of all the angles would be 360°.

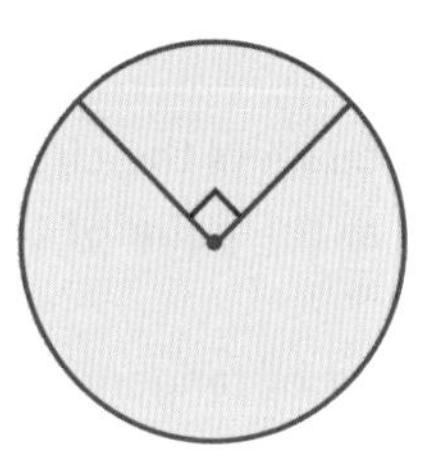

An **arc** is part of the circumference of a circle. We can measure an arc by its length (along the edge), or by the measure of its central angle.

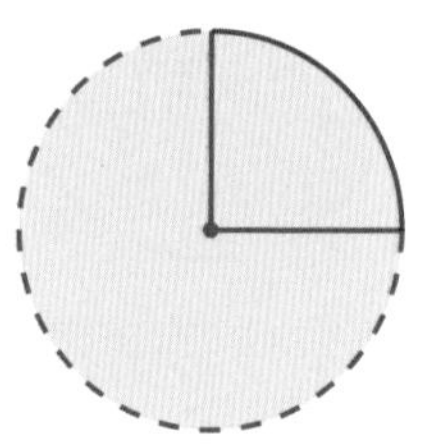

A **semicircle** is exactly half of a circle (sliced right through the center). A semicircle is a type of arc. Its degree measure is 180°.

Whenever an arc is defined in a circle, there are actually two arcs—the major arc and minor arc. A **major arc** has an angle measure greater than 180°, while a **minor arc** is less than 180°. In the figure below, the major arc has an angle of 265°, and the minor arc has an angle of 95°.

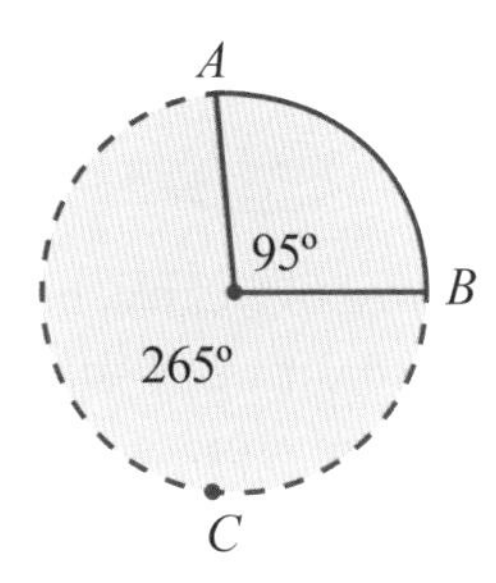

You can name a minor arc by its endpoints. To name a major arc, you should use three points. (This helps distinguish it from the corresponding minor arc.)

In the figure above, the minor arc could be named *AB*, while the major arc could be named *ACB*.

Note

A semicircle arc is neither a major arc nor a minor arc.

In a circle, congruent central angles have congruent arcs, and vice versa.

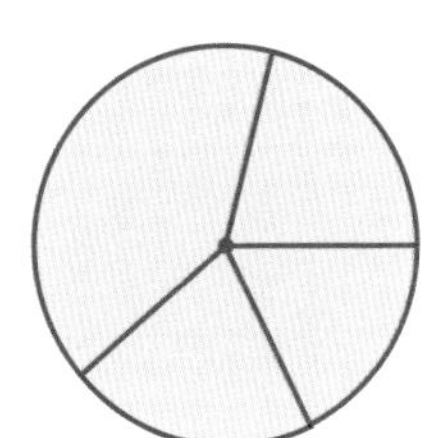

Furthermore, congruent chords have congruent arcs, and vice versa.

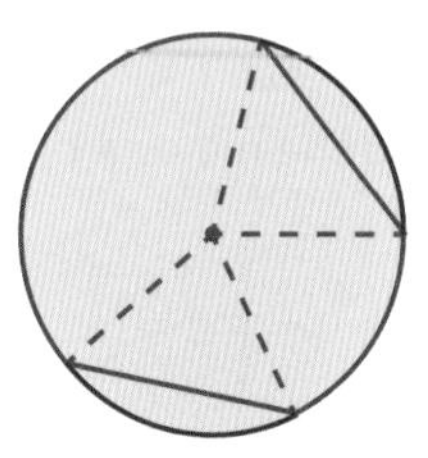

A **sector** is a portion of the area of a circle, bound by two radii and their intercepted arc. In other words, it's a "wedge," or "slice," of the circle. Sectors are measured in terms of area.

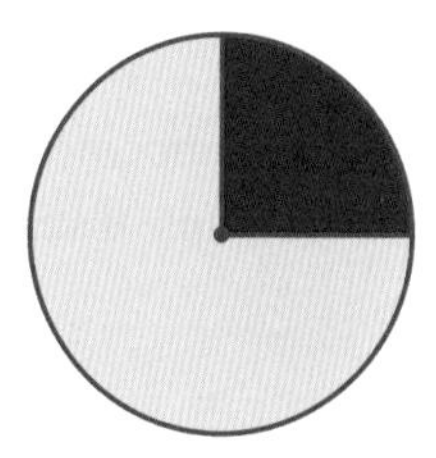

INSCRIBED ANGLES AND INTERCEPTED ARCS

An **inscribed angle** is an angle whose vertex is on the circumference of the circle. Both sides of the angle are chords.

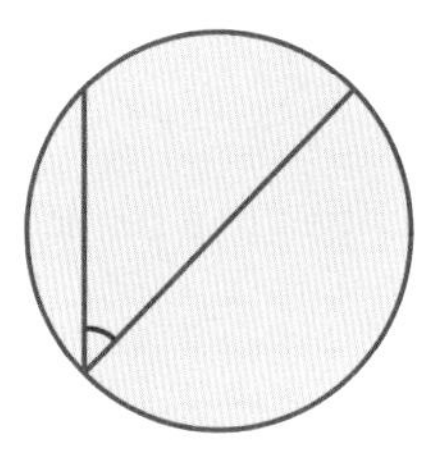

An **intercepted arc** is an arc that is defined by two lines or segments intersecting a circle. This is just a way of saying, "when we draw two lines intersecting a circle, an arc is formed." An intercepted arc is like any other arc, and its degree measure is found from the *central angle* that intersects those points.

You should be aware that an intercepted arc is considered to be the arc *between* two line segments. In the figure below, the intercepted arcs are shown in bold. Sometimes, there are two in the same circle.

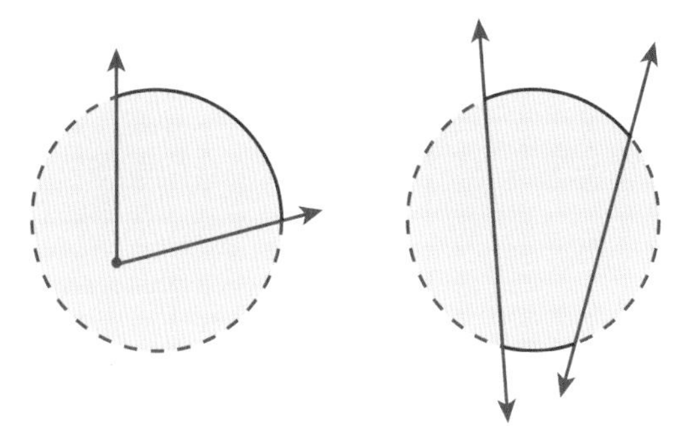

The measure of an inscribed angle is always half the measure of its intercepted arc.

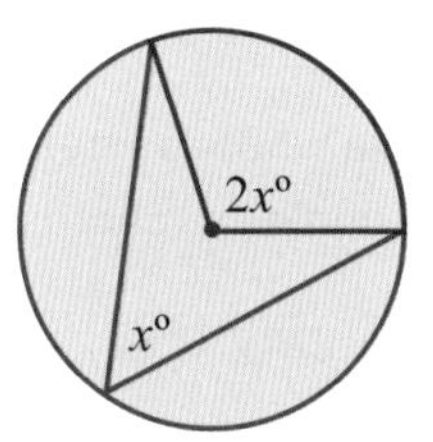

Therefore, if an angle is inscribed in a semicircle, then it is a right angle (90°).

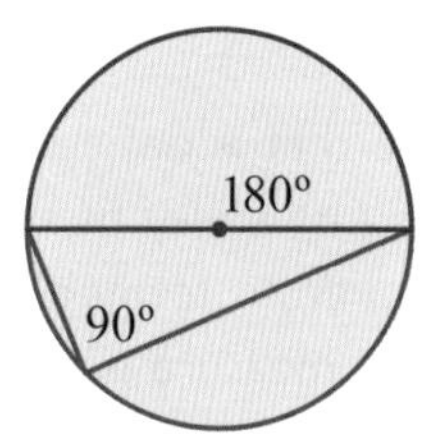

If two inscribed angles intercept the same arc, then the angles are congruent.

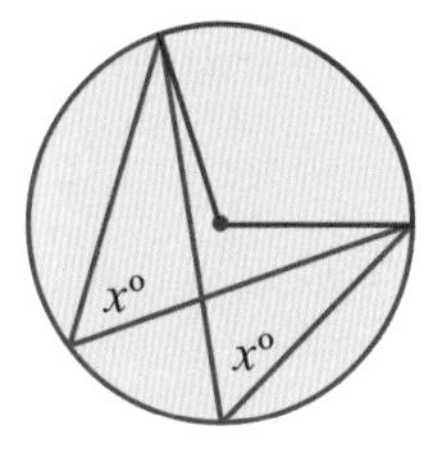

SLICES OF PI

In this section, you will learn how to calculate an arc/sector based on information about a circle, and vice versa.

Normally, the easiest way to think about these problems is in terms of basic fractions—that is, **part/whole** relationships.

Imagine that you have a pizza with 8 congruent slices. If you eat 1 of the slices, then you know that you ate $\frac{1}{8}$ of the total pizza.

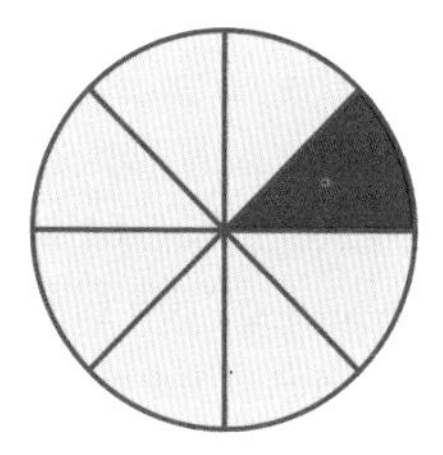

FINDING DEGREE MEASURE

Now, if you wanted to know the degree measure of the $\frac{1}{8}$ slice of pizza, that is also fairly straightforward. Since we know that the total degree measure of a circle is always 360°, we can use a proportion to find the angle of the arc. This is a part/whole relationship—the degree measure of the arc compared with the degree measure of the full circle.

$$\frac{1}{8} = \frac{x}{360^\circ}$$

Cross-multiply $$1 \times 360^\circ = 8 \times x$$

$$360^\circ = 8x$$

Divide both sides by 8 $$\frac{360^\circ}{8} = x$$

$$45^\circ = x$$

Therefore, the total degree measure of the $\frac{1}{8}$ slice is **45°**.

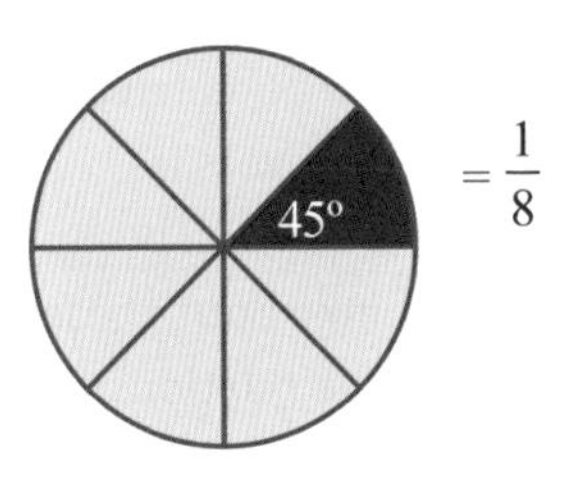

FINDING AREA

What if you wanted to know the total area of the $\frac{1}{8}$ slice you ate? To do that, you would use another part/whole relationship—the area of the slice compared with the total area of the pizza. So, you would need some information about the size of the pizza. Let's say that you know the radius of the circle is **12 inches**. That would be enough to tell you the area of the full pizza:

$$r = 12$$

$$A = \pi r^2$$

$$A = \pi(12)^2$$

$$A = 144\pi$$

Therefore, the area of the pizza is **144π** square inches. Now, we can use a proportion to find the area of the sector—that is, one slice of pizza.

$$\frac{1}{8} = \frac{s}{144\pi}$$

$$1 \times 144\pi = 8 \times s \qquad \text{Cross-multiply}$$

$$144\pi = 8s$$

$$\frac{144\pi}{8} = s \qquad \text{Divide both sides by 8}$$

$$18\pi = s$$

Therefore, the total area of the $\frac{1}{8}$ slice is **18π** square inches.

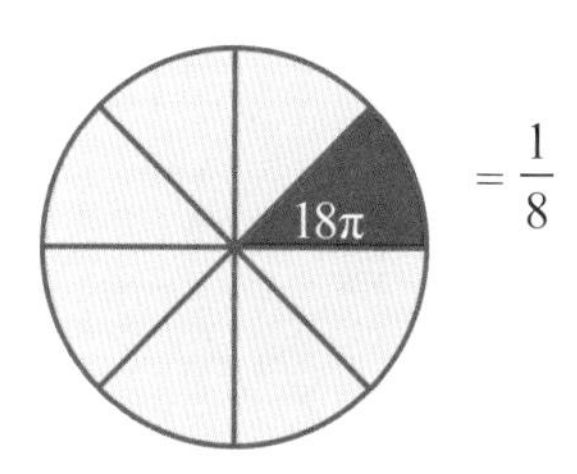

FINDING ARC LENGTH

Let's calculate the length of the arc (because obviously, the crust is the best part). To do that, we'll use another part/whole relationship—the length of the arc from the slice, compared with the total circumference of the pizza. We were told that the radius of the circle is 12 inches. With that, we can calculate the circumference of the pizza:

$$r = 12$$

$$C = 2\pi r$$

$$C = 2\pi(12)$$

$$C = 24\pi$$

Now, we can find the length of the arc:

$$\frac{1}{8} = \frac{a}{24\pi}$$

Cross-multiply $\quad 1 \times 24\pi = 8 \times a$

$$24\pi = 8a$$

Divide both sides by 8 $\quad \frac{24\pi}{8} = a$

$$3\pi = a$$

Therefore, the total arc length of the $\frac{1}{8}$ slice is **3π** square inches.

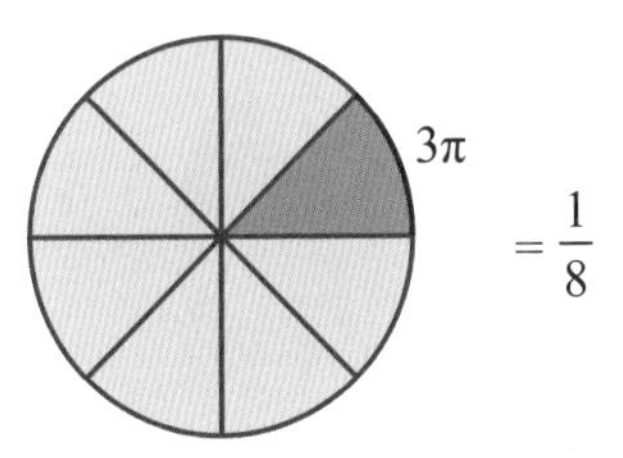

To solve a problem with arc or sectors, remember that all of the following are proportional:

Parts of a Circle

$$\frac{\text{part}}{\text{whole}} = \frac{\text{central angle}}{360°} = \frac{\text{arc length}}{\textit{total circumference } (2\pi r)} = \frac{\text{sector area}}{\textit{total area } (\pi r^2)}$$

You can solve for **all** of these parts if you have only two: something that tells you the size of the full circle, and something that tells you the size of the slice.

RADIANS

The **radian** is a unit of angle measure. One radian represents an angle in which the arc length is exactly equal to the radius of a circle.

The radian is usually described in terms of a circle with a radius of 1. In a circle with a radius of 1, one radian corresponds to an arc with a length of 1.

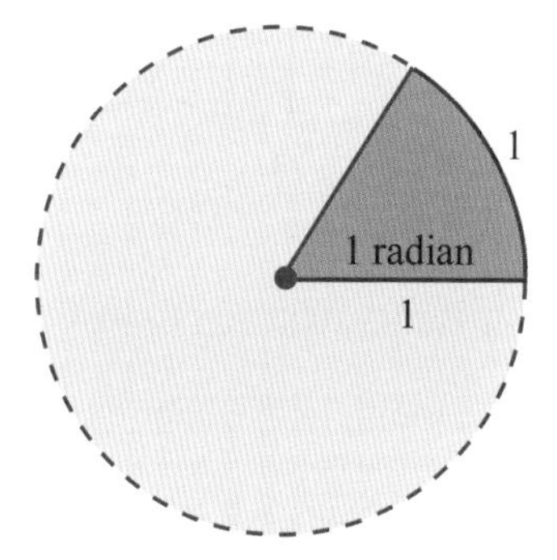

A circle has precisely **2π** (≈ 6.283) radians, a measure equal to **360°**.

Angles measured in radians are best expressed as fractions of 2π. For example, a half circle has an angle measure of π radians (= $\frac{1}{2}$ of 2π), and a quarter circle has an angle measure of $\frac{\pi}{2}$ radians (= $\frac{1}{4}$ of 2π).

Mathematicians like to use radians, because it is a unit based on a "real" relationship in a circle—the relationship between the arc and the radius. Degrees, by contrast, are actually arbitrary units. The reason we use 360° is because 360 has a lot of factors, which makes division easy.

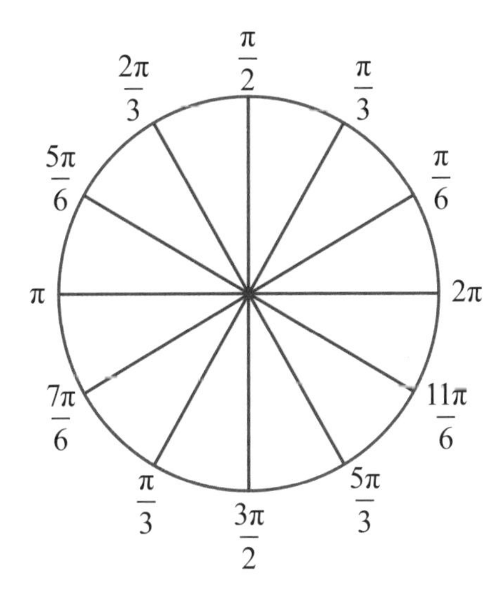

If the figure is confusing, it may help you to write down the unreduced form of each fraction as a multiple of $\frac{1}{6}$. For example, $\frac{\pi}{3}$ is equal to $\frac{2\pi}{6}$, and so on. You'll see that the values are sequential: $\frac{\pi}{6}, \frac{2\pi}{6}, \frac{3\pi}{6}$, etc.

> If you see an angle measure expressed in terms of π, you can be sure that you're dealing with radians. Sometimes, the abbreviation "r" is used, but it is not necessary.

To convert radians to degrees, or vice versa, remember that **π radians = 180 degrees**.

You can also use the following proportion:

Converting Radians and Degrees

$$\frac{\text{radians}}{\pi} = \frac{\text{degrees}}{180}$$

Arcs and Sectors Using Radians

$$\frac{\text{part}}{\text{whole}} = \frac{\text{central angle}}{2\pi} = \frac{\text{arc length}}{\text{total circumference } (2\pi r)} = \frac{\text{sector area}}{\text{total area } (\pi r^2)}$$

Here is another formula that may be useful for working with circles and radians.

Arc Length, Using Radians

$$s = \alpha r$$

where r is the radius, s is the length of the arc, and α is the central angle measure in radians.

Note that the Greek letter α (alpha) is commonly used to represent an angle measure in radians. Another frequently used variable symbol is θ (theta).

INCIRCLE AND CIRCUMCIRCLE

The **incircle** of a polygon is a circle that is drawn inside the polygon and is **tangent** to every side. Another term for incircle is **inscribed circle**. The center of an incircle is called the **incenter**.

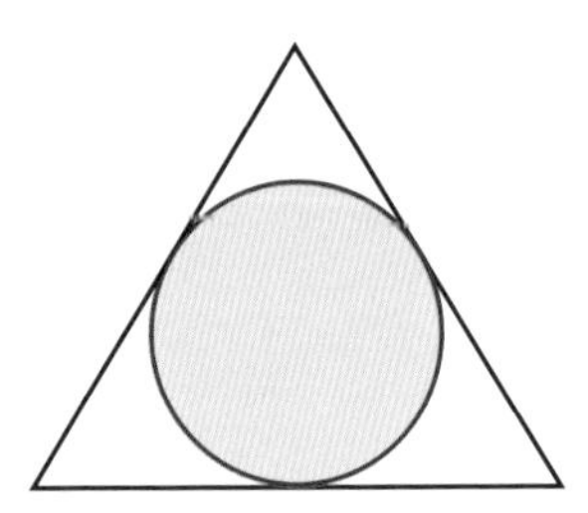

All triangles have incircles, as do all regular polygons. Irregular polygons with more than three sides will not necessarily have an incircle that touches every side.

Some sources use a looser definition of incircle which can apply to all irregular polygons: "An incircle is the largest circle that can be drawn completely inside a given polygon."

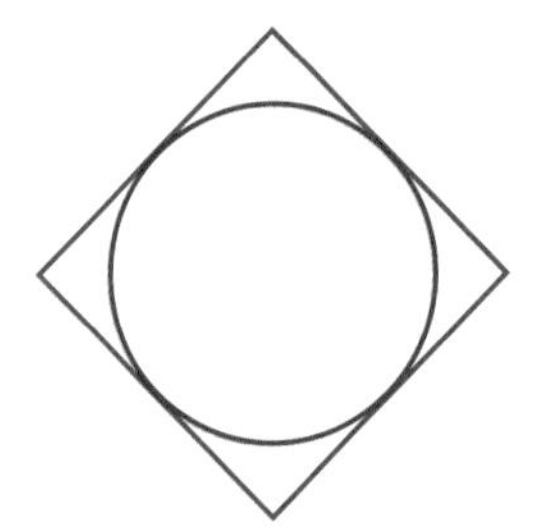

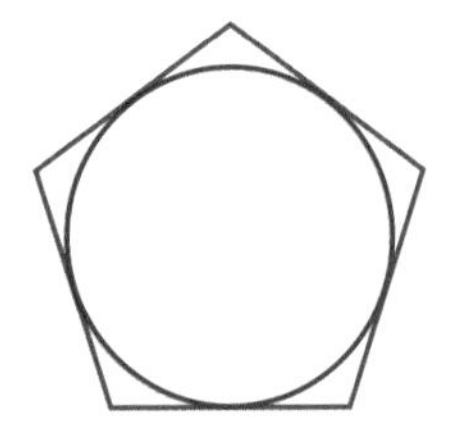

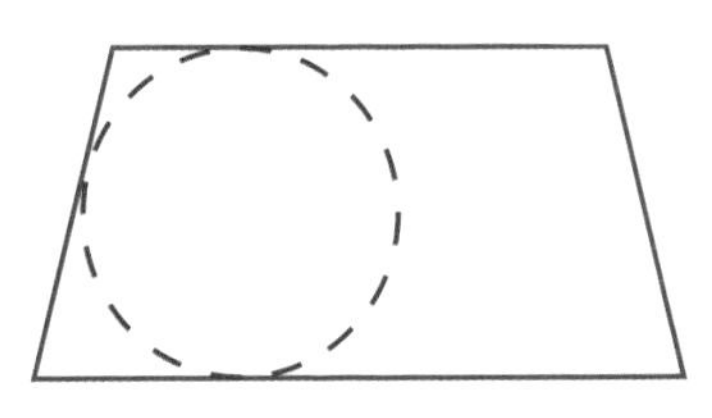

The two polygons on the left have an incircle, while the one at the right does not.

The radius of the incircle is known as the **apothem** or **inradius**. An apothem segment connects the incenter to a given side of the polygon, and is also perpendicular to that side. If a polygon has an incircle, then each side of the polygon has one apothem.

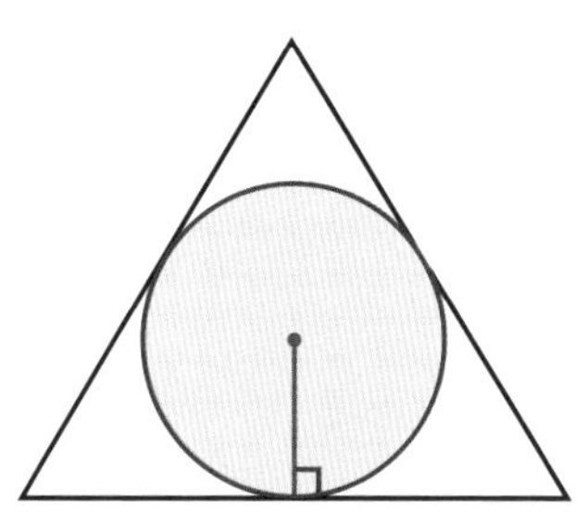

The **circumcircle** of a polygon is a circle that is drawn outside the polygon and touches every vertex. Another term for circumcircle is **circumscribed circle**. The center of a circumcircle is called the **circumcenter**.

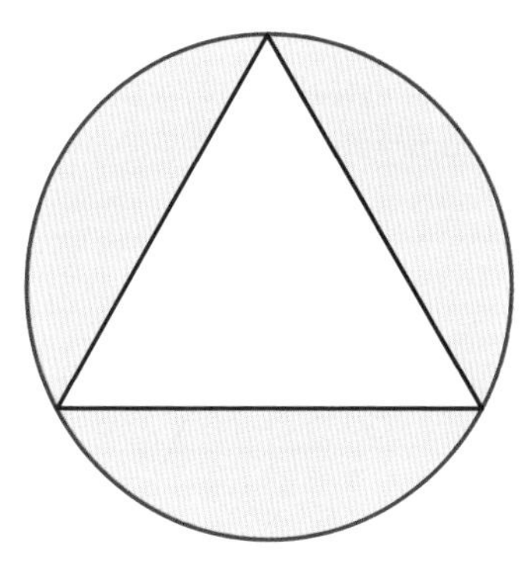

All triangles have circumcircles, as do all regular polygons. Irregular polygons with more than three sides will not necessarily have a circumcircle that touches every side.

> Some sources use a looser definition of "circumcircle," which can apply to all irregular polygons: "A circumcircle is the smallest circle that can be drawn completely outside a given polygon."

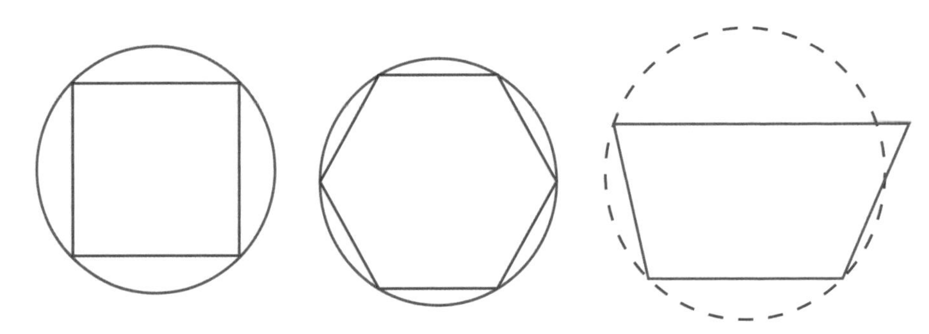

The two polygons on the left have a circumcircle, while the one at the right does not.

The radius of the circumcircle is known as the **circumradius** or just **radius.** A circumradius segment connects the circumcenter to a given vertex of the polygon. If a polygon has a circumcircle, then each vertex of the polygon has a circumradius.

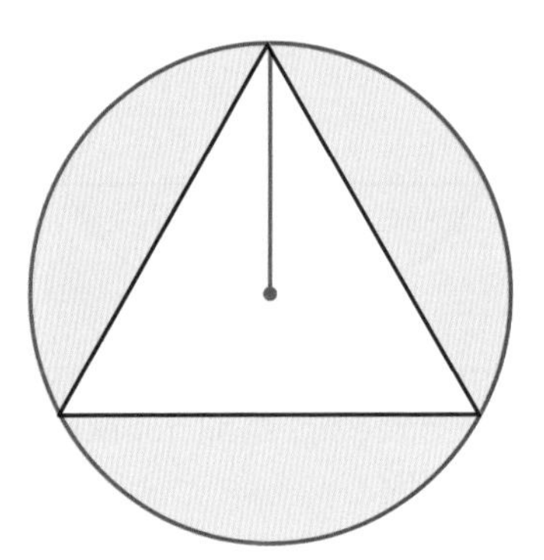

CHAPTER

5 CONGRUENCE AND SIMILARITY

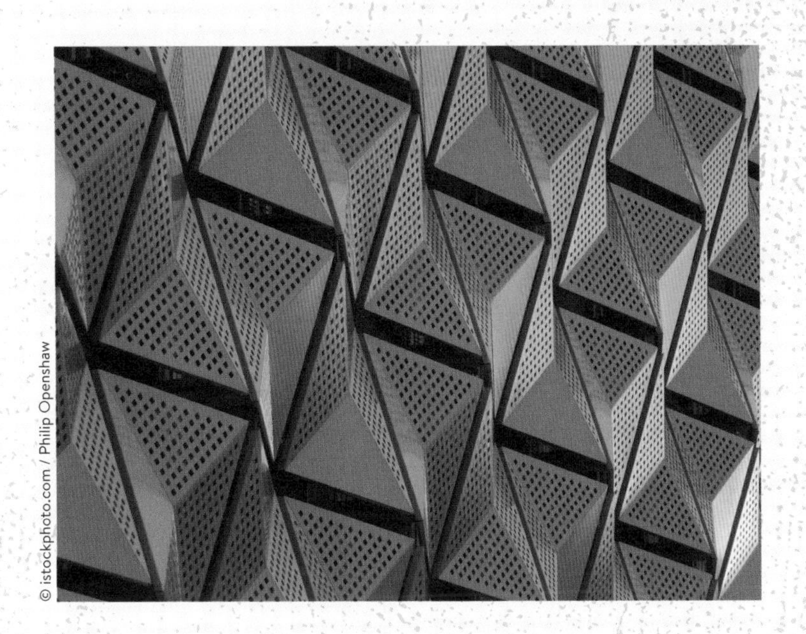

Congruence and similarity are crucial concepts in geometry. What's more, showing that two figures have one of these relationships will often be easier than you might think.

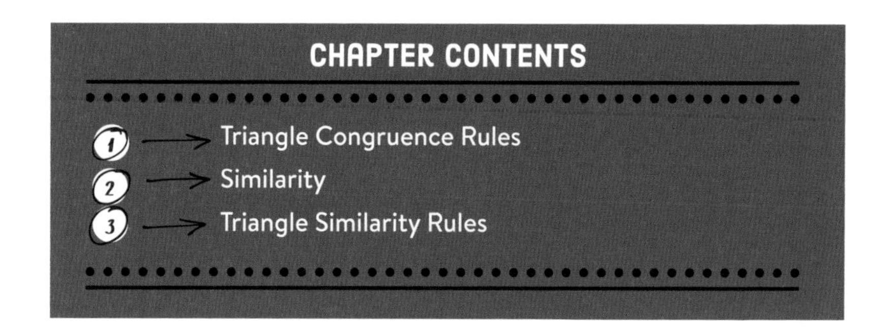

Congruence is an important concept in Geometry. In fact, a great number of geometric rules and theorems are based on congruence. In this chapter, we will review several of these rules.

To recap, if two figures are congruent, it means that they have the same shape and size. If two polygons are congruent, it means that all of their side lengths and angles are congruent, too.

In geometry, **congruent** means *having exactly the same shape and size.*

- Congruent line segments have exactly the same length.
- Congruent angles have exactly the same degree measure.
- Congruent shapes have the same side lengths and angle measures.

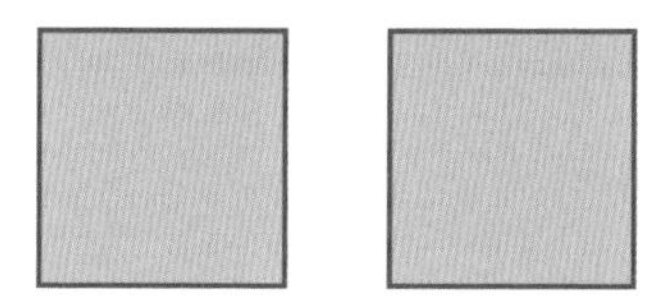

Example of congruent figures

TRIANGLE CONGRUENCE RULES

In the case of triangles, we can often derive a lot of information when we know even just a couple of facts (e.g., angle measures or side lengths). In this section, we'll review several basic theorems regarding congruent triangles.

First, if two triangles have the same side lengths, then the triangles are congruent.

SSS (Side-Side-Side) Congruence Postulate

If the three sides of one triangle are congruent to the three sides of another triangle, then the two triangles are congruent.

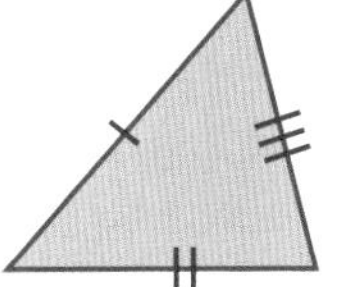

Try it yourself! You can demonstrate this postulate using physical objects, such as straws, to form the sides of the triangle.

Recall that if two angle measures are known, then we can find the third angle as well (the sum of the three angles is always 180°). Furthermore, if all three angles are known, then we know the triangle's shape. If we also specify a side length, then we know the triangle's size.

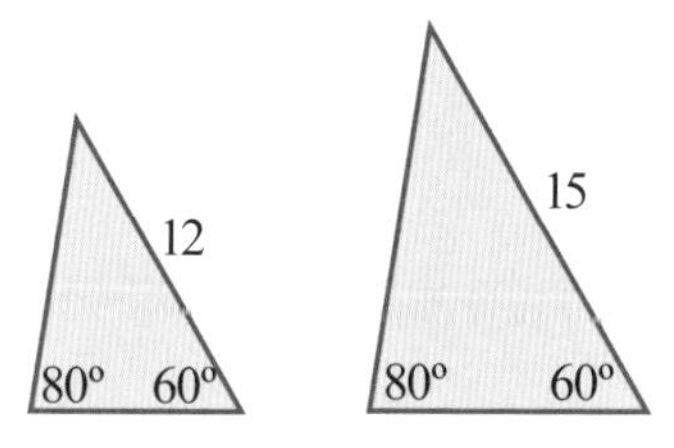

Therefore, we have two congruence postulates that we can use when triangles have two (actually three!) angle measures and one side length in common.

ASA (Angle-Side-Angle) Congruence Postulate

If two angles and the included side of one triangle are congruent to the corresponding parts in another triangle, then the two triangles are congruent.

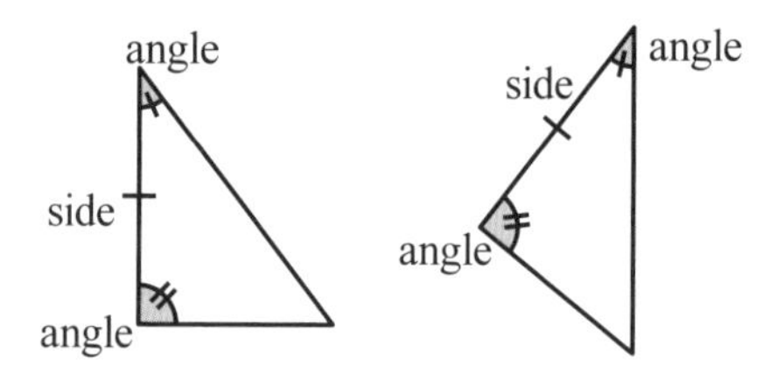

For these postulates, **included** means it's contained between two other parts. An **included side** is the side between two specified angles. An **included angle** is the angle between two specified sides.

Congruent angles or sides are often identified with hash marks like these. In this figure, the angles marked with a single hash are a congruent pair, and the ones with a double hash are a different congruent pair.

AAS (Angle-Angle-Side) Congruence Postulate

If two angles and a non-included side of one triangle are congruent to the corresponding parts in another triangle, then the two triangles are congruent.

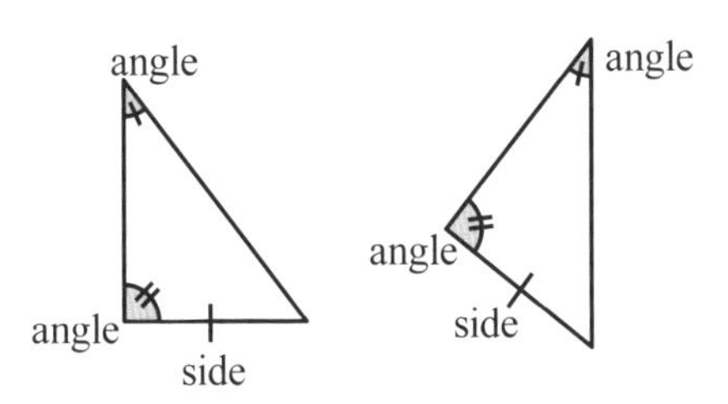

There's an important catch that you must remember—the known side must be in the same place, relative to the angles, in both triangles. That is, if the known side is opposite to the smallest angle in the first figure, then that must also be true for the second figure.

Another way to think of this is that order matters—as you go around the triangle in either direction, you have angle, angle, and side, (or side, angle, and angle), one after the other, in that order. If the second triangle has those same parts with the same measurements in the same order, then the two triangles are congruent.

SAS (Side-Angle-Side) Congruence Postulate

If two sides and the included angle of one triangle are congruent to the corresponding parts in another triangle, then the two triangles are congruent.

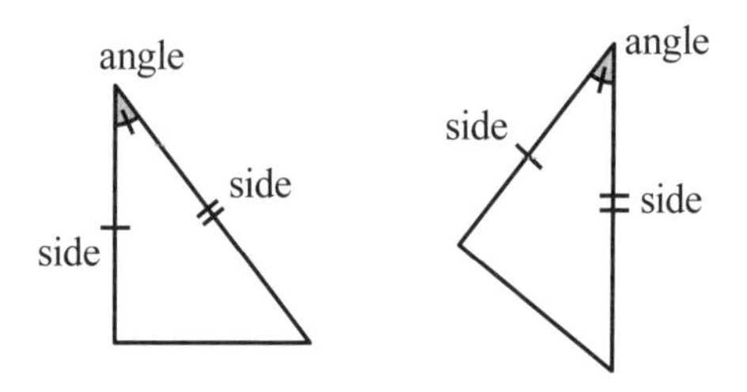

Again, order matters with this postulate, as it does with the ASA and AAS postulates.

Additionally, you may wonder if there is an SSA (Side-Side-Angle) postulate. This particular combination does *not* work for triangle congruence, so there is **no SSA postulate**. The reason is that if you have two side lengths and a non-included angle specified, then there are exactly two triangles that can exist with that combination.

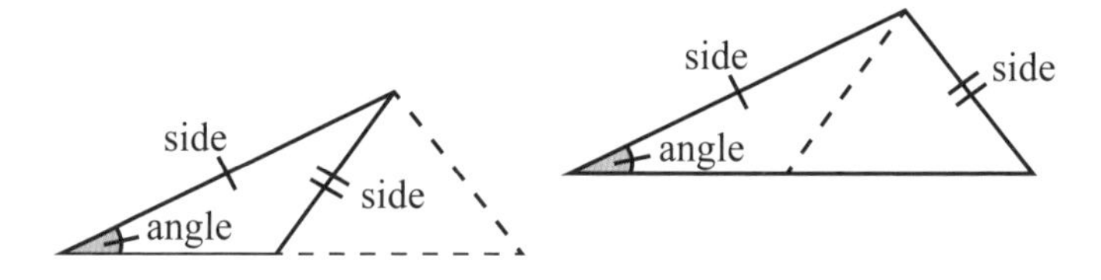

In the example above, there are two different ways the side with two hash marks can fit into the space created by the marked angle.

Furthermore, there is not a congruence postulate for AAA (Angle-Angle-Angle). You'll learn about **similarity** postulates in the next section.

SIMILARITY

Similar figures have the same shape, but not necessarily the same size. If two polygons are similar, then they have all the same angle measures in corresponding positions. Additionally, their side lengths are proportional.

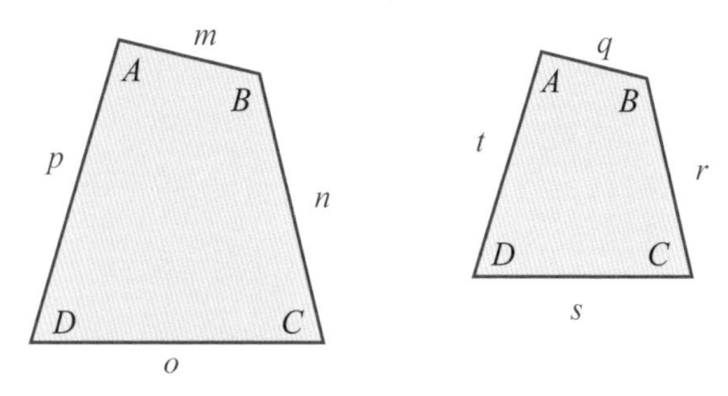

In the figure above, if the figures are similar, then $\angle A = \angle A$, $\angle B = \angle B$, and so on. Also, the corresponding pairs of sides are proportional. For example, $\frac{m}{q} = \frac{n}{r}$, $\frac{o}{s} = \frac{p}{t}$, and so on.

The **scale factor** of a pair of similar figures is the ratio of their corresponding side lengths. For example, if you enlarge a figure by doubling its side lengths, then the scale factor of the figures is 2.

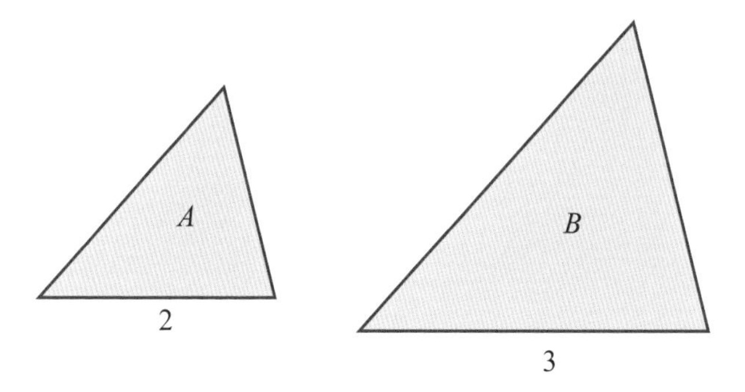

In the figure above, the scale factor of triangle *A* to triangle *B* is 2:3.

TRIANGLE SIMILARITY RULES

SSS (Side-Side-Side) Similarity Postulate

Two triangles are similar if all three pairs of corresponding sides are proportional.

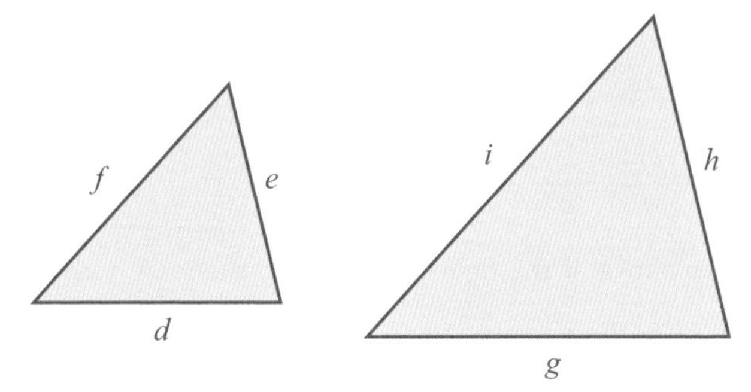

In the figure above, if $\frac{d}{g} = \frac{e}{h} = \frac{f}{i}$, then the two triangles are similar.

AAA (Angle-Angle-Angle) Similarity Postulate

Two triangles are similar if all three pairs of corresponding angles are congruent.

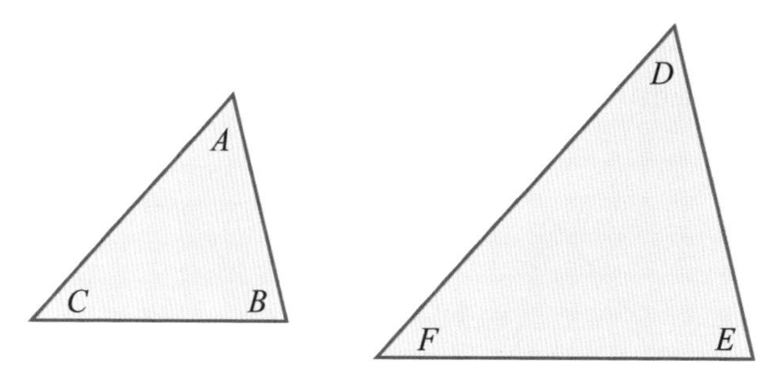

In the figure above, if $\angle A \cong \angle D$, $\angle B \cong \angle E$, and $\angle C \cong \angle F$, then the two triangles are similar.

As a matter of fact, we need to verify only *two* pairs of corresponding angles. Recall that for any triangle, if we know two of the angle measures, then we know the third angle measure as well. For this reason, the AAA Similarity Postulate is also known as the **AA Similarity Postulate**.

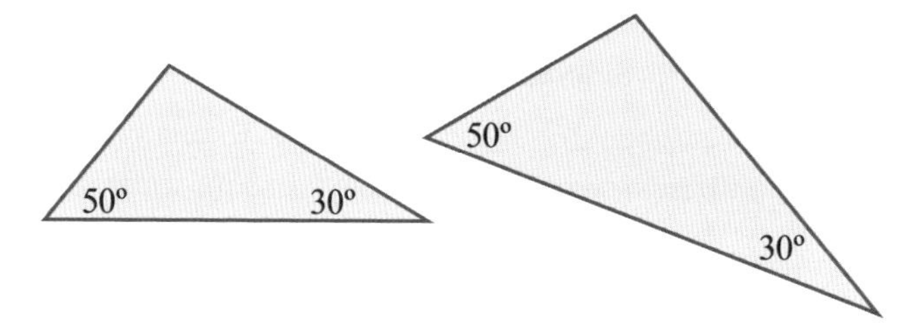

In the figure above, two pairs of angles are known. Can you figure out the measure of the remaining two angles? That's right—they're both the same!

SAS (Side-Angle-Side) Similarity Postulate

If two sides and the included angle of one triangle are similar to the corresponding parts in another triangle, then the two triangles are similar.

That is to say, the two pairs of corresponding sides are proportional, and the pair of included angles are congruent.

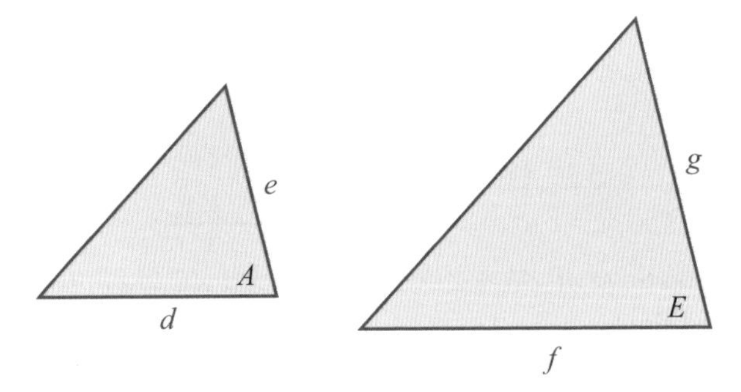

In the figure above, if $\frac{d}{e} = \frac{f}{g}$ and $\angle A = \angle E$, then the two triangles are similar.

Side-Splitter Theorem

A line drawn through a triangle, if parallel to one of the triangle's bases, divides the triangle into similar parts.

In other words, if you have an inscribed triangle that has a base parallel to the original triangle, then the two triangles are similar.

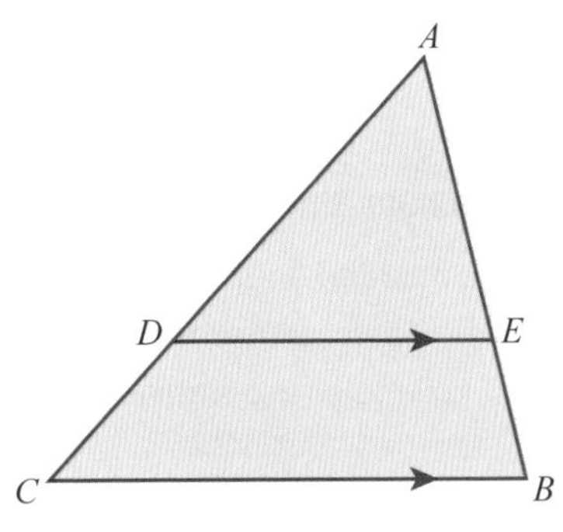

In the figure above, triangle *ABC* is similar to triangle *AED*.

SIMILARITY IN RIGHT TRIANGLES

Of course, all of the rules discussed in this chapter apply to right triangles as well. Here is another rule that is useful for right triangles:

Right Triangles and their Altitudes

The altitude of a right triangle (perpendicular to the hypotenuse) divides the triangle into similar parts. The two smaller triangles are similar to each other and to the larger triangle.

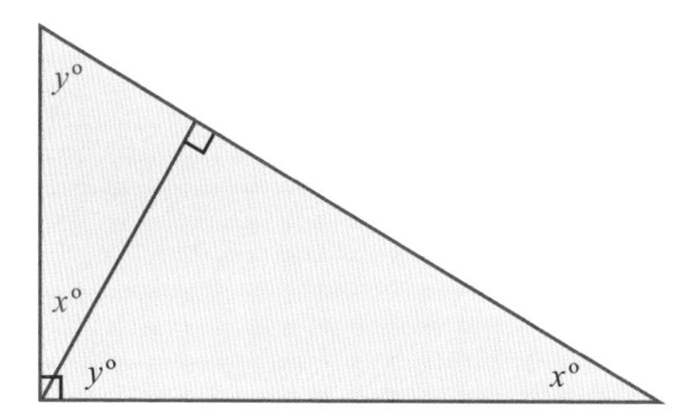

This is because the triangles have congruent angle measures. That is, if the original triangle has angles *x* and *y*, then the altitude divides the right angle into *x* and *y* as well.

ASK YOURSELF...

Consider what you know about triangles and complementary angles. Do you understand why the rule above is true?

CHAPTER

CONSTRUCTIONS

© istockphoto.com / bestdesigns

Ancient peoples obviously didn't have the advanced machines that we use today. Although simple drawing programs allow us to create figures such as perfect circles or regular heptagons with a few clicks of a mouse, the ancient world could rely on only a compass and a straightedge. Even with today's machines, mathematicians are fascinated by what can be done with the ancient tools.

CHAPTER CONTENTS

CONSTRUCTING LINES AND ANGLES

These days, if we want to create a perfectly precise image with circles, straight lines, or other shapes, we could just use a computer program. However, throughout history, when people needed to be precise, they had to get creative by using instruments. A **compass** is an instrument that aids in drawing perfect circles of almost any size. A **straightedge** aids in drawing perfectly straight line segments. You'll most likely use a ruler as your straightedge. However, keep in mind that when drawing compass-and-straightedge constructions, you should not be using measuring tools, such as the markings on your ruler. You can find compasses and rulers in the school supply section of any well-stocked department store.

Supplies
For this section, you should have your compass, straightedge, and scratch paper ready. You should be actively practicing the constructions as you read.

Geometry classes today still ask students to use compass-and-straightedge constructions. This is actually a great way to gain a hands-on understanding of concepts such as bisector, perpendicular, and congruence.

CONSTRUCTING LINE SEGMENTS

The first basic exercise is to draw a straight line through two points. This may seem trivial to accomplish; however, some people have trouble getting things lined up exactly right, which is mostly due to the thickness of a typical ruler. Practice getting your line to cross the points precisely. Keep your pencil sharp for these exercises, and angle the pencil lead so that there's no gap between the lead and the edge of the ruler.

Some compasses are cheaply made and frustrating to use. A good compass will not have wobbly parts and will hold its radius and pencil lead quite firmly when you use it. If you're having a hard time making circles, the problem might be your compass, not you!

PERPENDICULAR LINE SEGMENTS

In this exercise, you'll use your compass and straightedge to construct two line segments that are precisely perpendicular to each other. If you've never used a compass before, it may take some practice to get the hang of it. If needed, try making some practice circles on your own first.

METHOD 1–START WITH A POINT

A

It's important to label points where indicated in the instructions because you're going to connect these points later.

Begin with a point, as in the figure above. For the purposes of this exercise, label the point A. Position the compass so that the needle is on point A, and make a circle that is perhaps a couple of inches in diameter.

Note that with any of these constructions, it's generally better to use larger circles when possible, as ultra-small circles make it more difficult to find precise intersection points. Strike a balance between reasonable size and clarity in your figures.

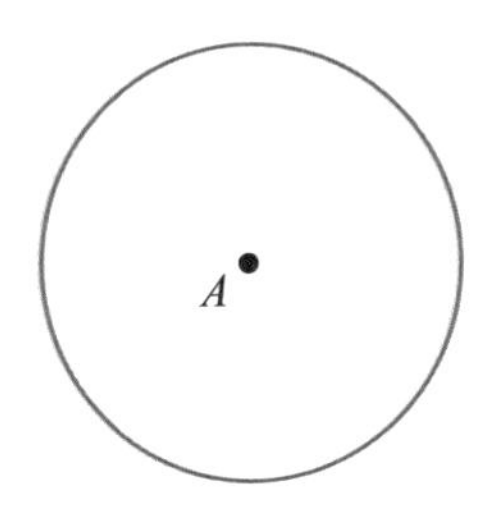

Now, choose any point on the circumference of the circle, and label it point B.

Note: you could have also started with two points, A and B. You would make a circle by placing the compass needle on A and the drawing point on B. Either way, at this step you'll have a figure like the one below:

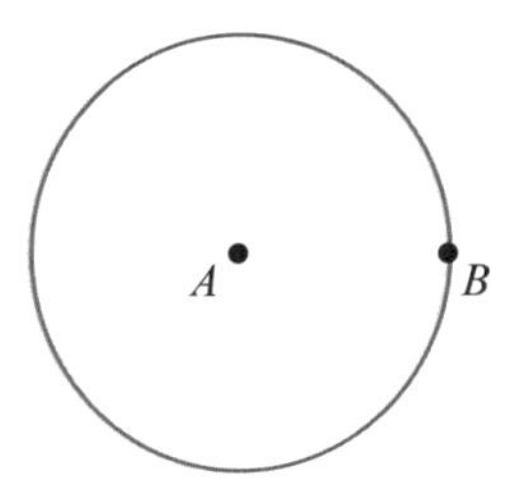

Next, position the compass so that the needle is on point B, and the drawing point is on A. Try to be precise! With this radius, make a second circle which will intersect the first circle in exactly two places: points C and D. Note that this circle has exactly the same radius as circle A.

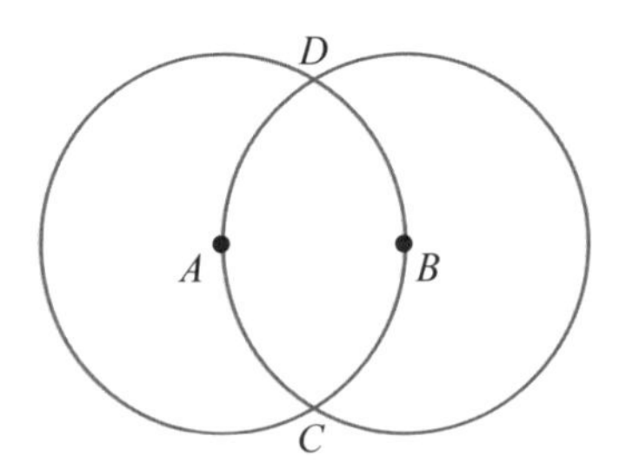

Finally, use your straightedge to draw a line that connects points *A* and *B*. Also draw a line that connects the two intersection points of the circles (points *C* and *D* in the figure above).

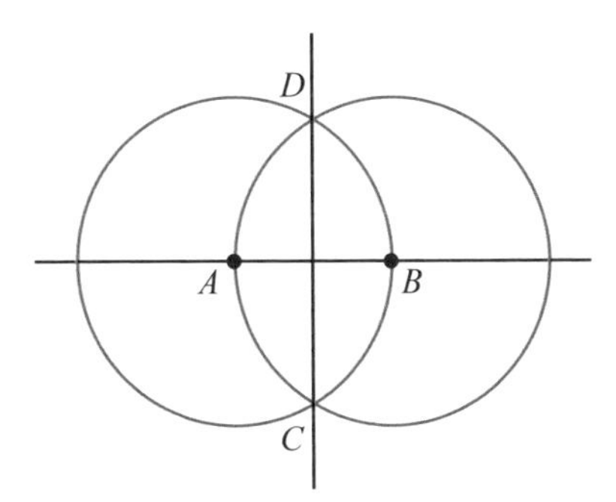

The two lines are precisely perpendicular to each other. One reason this works is that each pair of adjacent points is the same distance apart, as shown in the figure below. Those four points would actually form a **rhombus**, and by definition, a rhombus has diagonals that are perpendicular to each other.

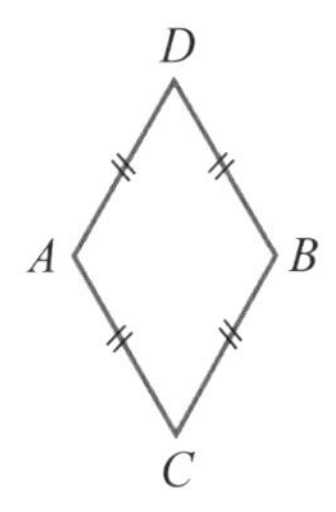

METHOD 2–START WITH A LINE SEGMENT

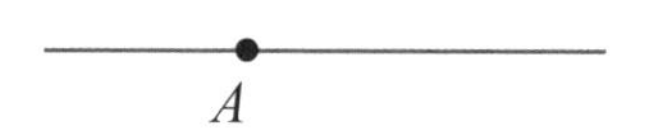

This method is essentially the same as Method 1, except that we begin with a line segment instead of just a point. Begin with a straight line segment like the one above, with a point labeled *A* on the line segment. Using your compass, make a circle centered at point *A*.

This circle intersects the line segment in exactly two places. Choose either one of these intersection points and label it *B*.

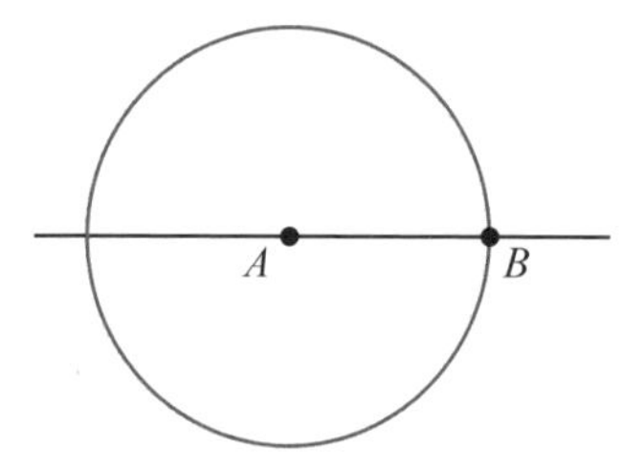

Next, position the compass so that the needle is on point *B*, and the drawing point is on *A*. With this radius, make the second circle, which will intersect the first circle at exactly two places: label the points *C* and *D*.

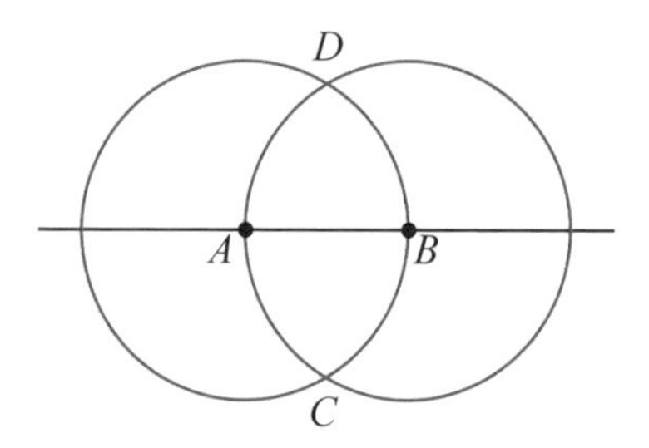

Finally, use your straightedge to draw a line between those two intersection points.

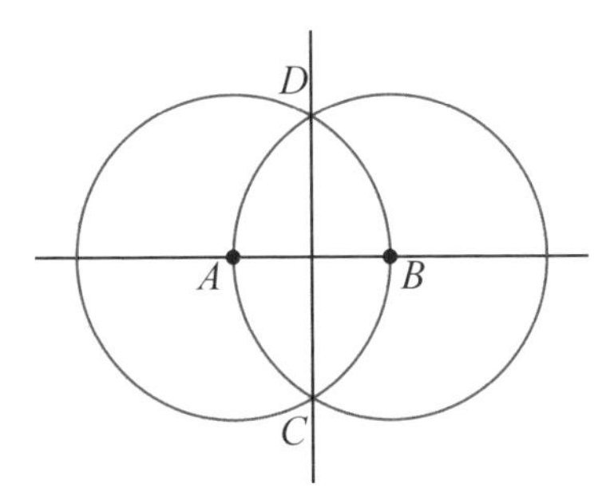

The two lines are precisely perpendicular to each other.

Once you're comfortable with these constructions, you may be able to avoid drawing complete circles with the compass, but instead draw partial arcs at the predicted point(s) of intersection. You would follow the same steps above, but eyeball the figure and draw arcs only at the relevant spots on the figure. This is a totally optional adjustment, and may take practice.

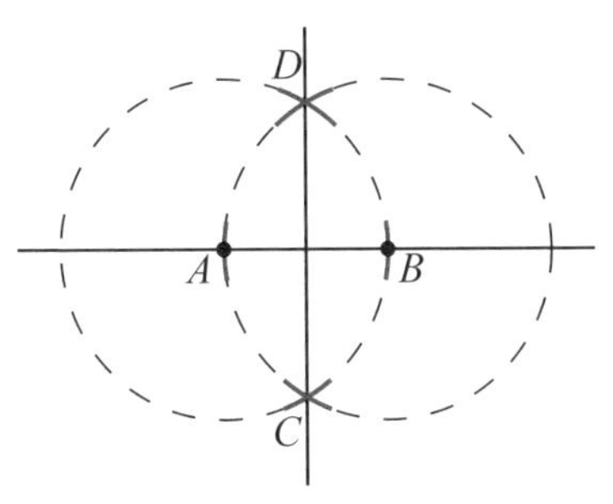

From here on out, we'll show the full circles with dashes and the relevant arc spots highlighted in **red**.

Try a few more for practice! Follow the steps above to construct perpendicular lines. Repeat until you're comfortable with the process.

PERPENDICULAR BISECTOR OF A LINE SEGMENT

Next, you'll use your compass to construct a **perpendicular bisector** of a line segment. This is the same as the previous exercise, but you'll use the endpoints of the line segment (*A* and *B* in the figure) as your compass points. Use this approach when you need your new perpendicular line to intersect at the exact midpoint of the first line. Begin with a line segment like the one above, with the endpoints labeled *A* and *B*.

Use the same strategy from the previous sections. Construct two circles—one centered at *A* and intersecting *B*, and vice versa for the other circle.

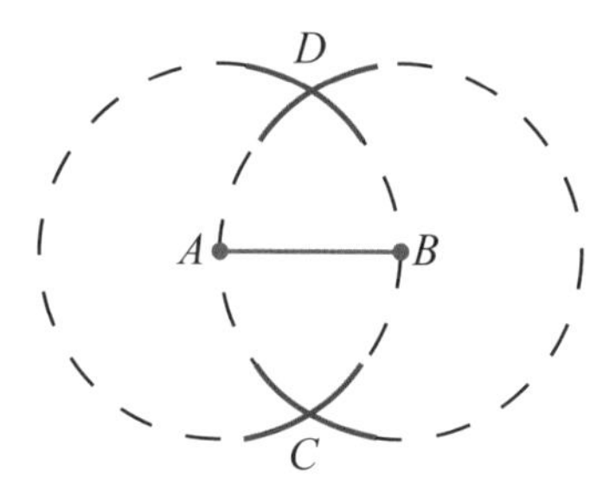

Finally, use your straightedge to draw a line between those two intersection points.

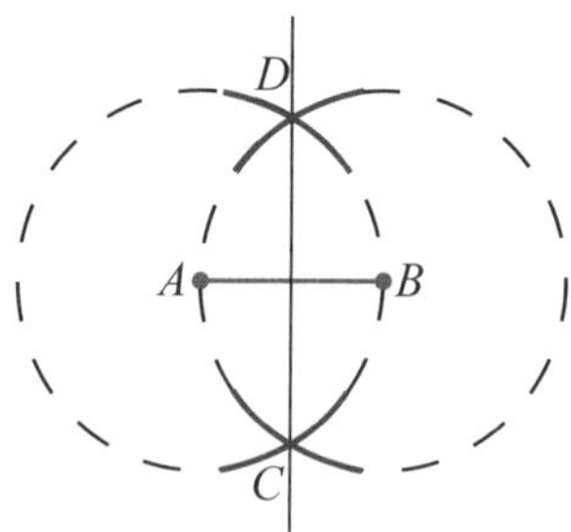

The two lines are precisely perpendicular to each other. Additionally, the new perpendicular line intersects the exact midpoint of the line segment *AB*.

Try a few more for practice! Follow the steps above to create a perpendicular bisector of a line segment. Repeat until you're comfortable with the process.

PERPENDICULAR LINE THROUGH A POINT

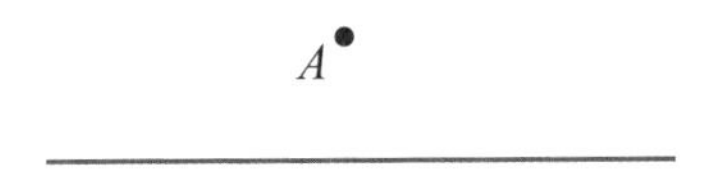

Next, you'll use your compass to construct a line that is perpendicular to an existing segment and also passes through a given point. This is a little like the perpendicular bisector construction, but in reverse. Use this approach whenever you need a perpendicular line to go through a particular point in the figure.

Begin with a line or segment like the one above, and a point *A* **not** on the line. Position the compass so that the needle is on *A*, and make a circle that intersects the line in two places, points *C* and *B*.

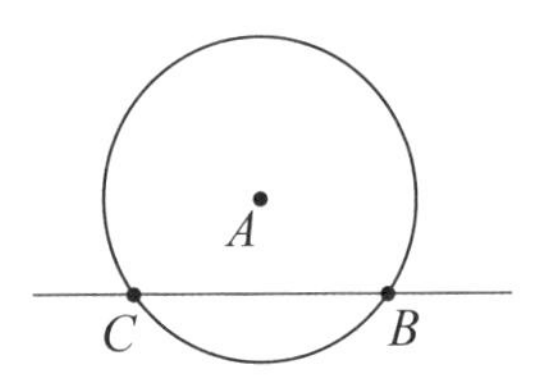

Then, place the compass needle point on *B* and the drawing point on *A*. Make a circle.

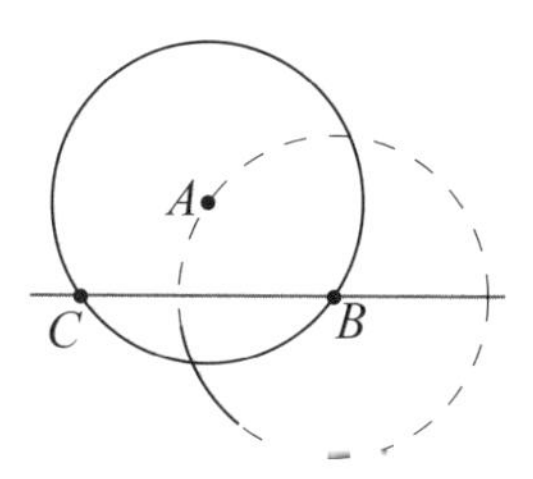

Then, place the compass needle point on *C* and the drawing point on *A*. Make a circle.

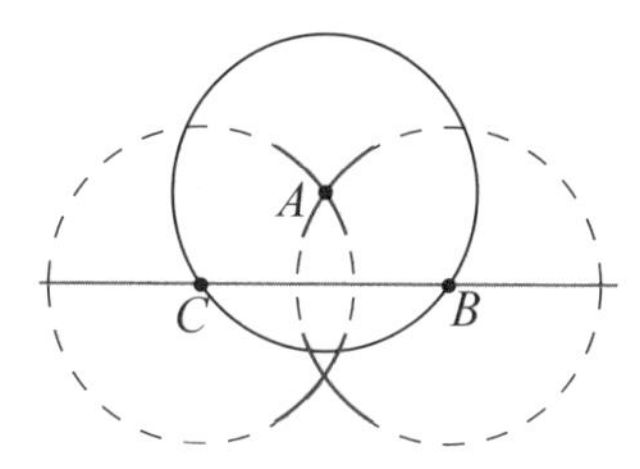

Finally, use your straightedge to draw a line between those two intersection points.

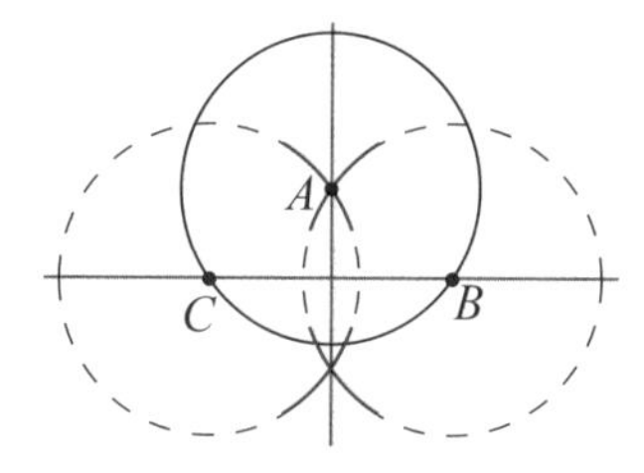

The two lines are precisely perpendicular to each other, and they pass through point *A*.

Try a few more for practice! Follow the steps above to create a perpendicular line through a point. Repeat until you're comfortable with the process.

ANGLE BISECTOR

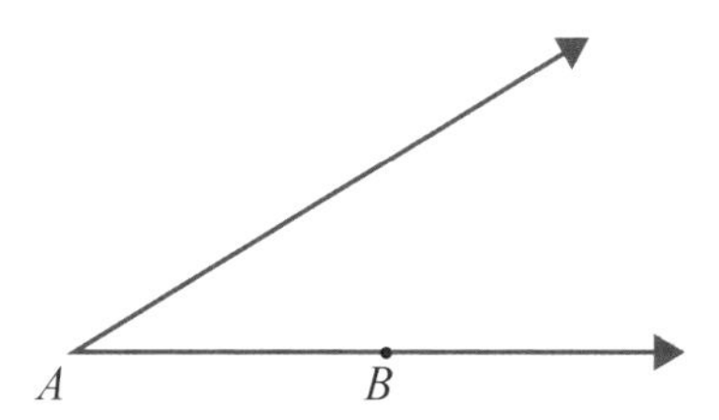

In this exercise, you'll construct a line segment that precisely **bisects** an existing angle. Begin with an angle like the one above (drawn using a straightedge), with the vertex labeled *A* and another point labeled *B* on one of the angle's legs. Position your compass so that the needle point is on *A*, and the drawing point is on point *B*. Turn the compass to make a circle, or just an arc that crosses both of the angle's legs. Each of these two intersection points will be equidistant from the angle's vertex. Label other intersection point *C*.

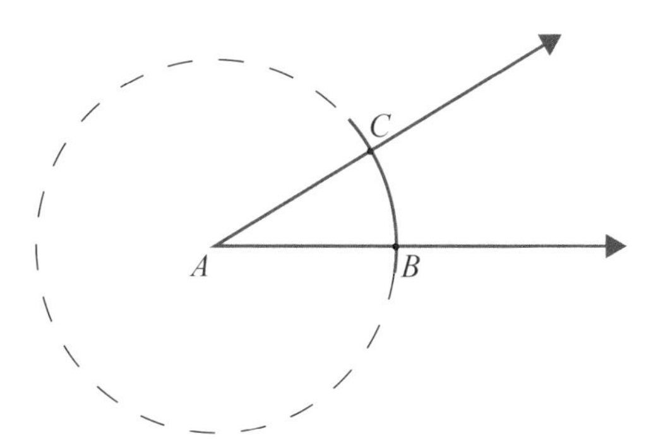

Now, position the compass needle point on *B* and the drawing point on the vertex *A*. With this radius, make a circle or just an arc that spans the angle's interior.

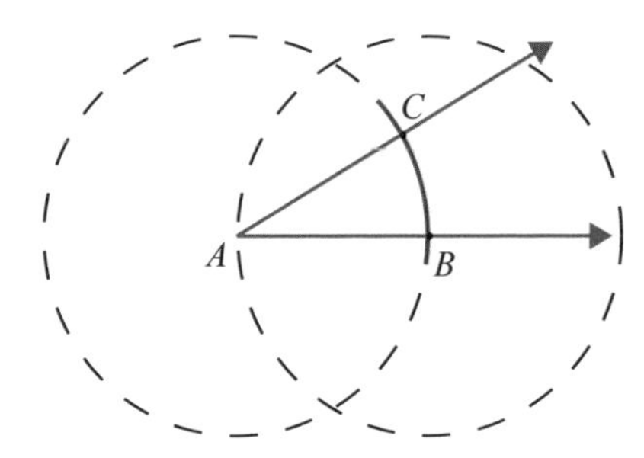

Then, position your compass with the needle point on *C* and the drawing point on vertex *A*. With this radius, make a circle or just an arc that spans the angle's interior.

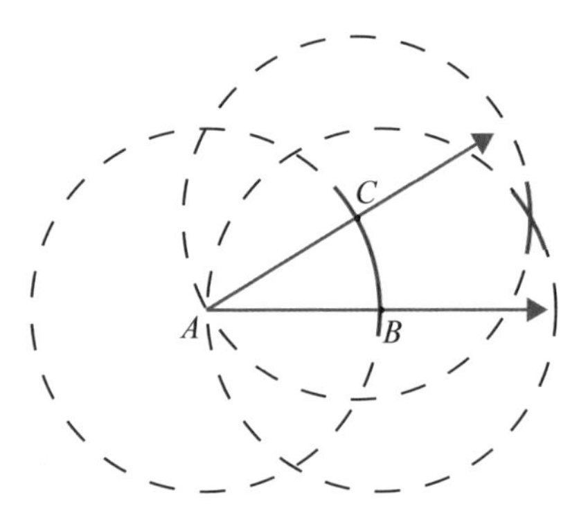

Finally, use your straightedge to draw a line from this new intersection point (labeled *D* in the figure below) to the angle's vertex.

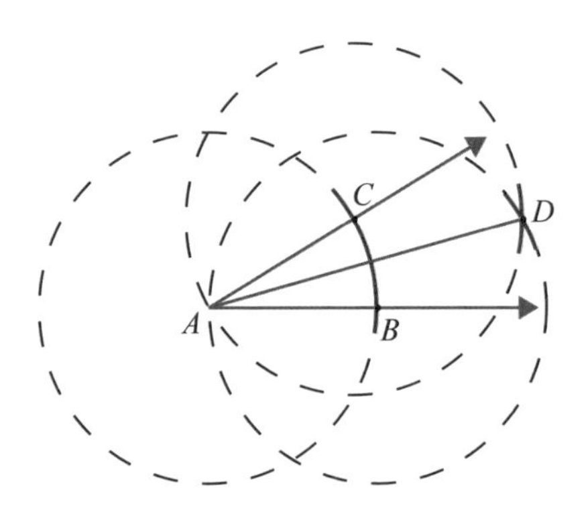

This new line is a bisector of the original angle—it divides the angle precisely in half. One way to prove this is with the SSS triangle congruence theorem. If we connect points to create triangles *ABD* and *ACD*, these triangles would be congruent because they have three pairs of corresponding congruent sides.

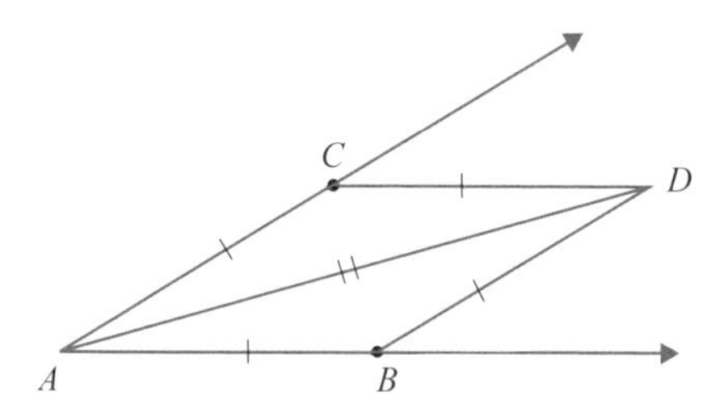

Try a few more for practice! Follow the steps above to construct an angle bisector. Repeat until you're comfortable with the process.

COPY AN ANGLE

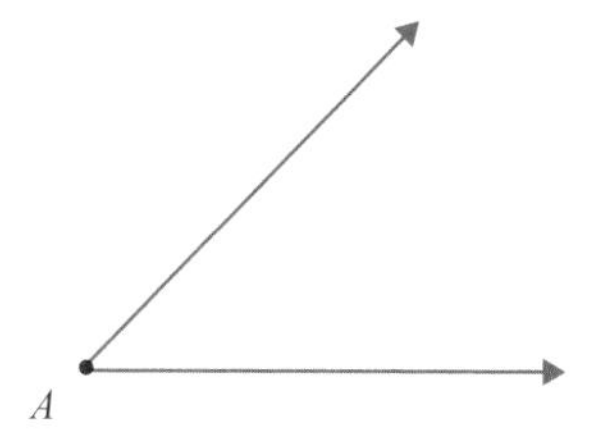

Here, you'll construct an angle that is congruent to another angle. Begin with an angle like the one above, with the vertex labeled *A*. Then, use your straightedge to draw another line segment nearby. This line segment will be one of the legs of the copied angle. Draw a point on the line segment, and label it *A'*.

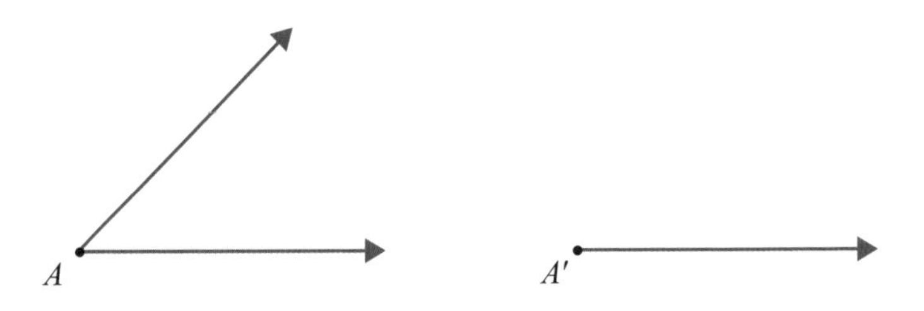

Next, position the compass needle point on vertex *A*, and make an arc that intersects both of the angle's legs. The arc can be any size. Label the intersection points *B* and *C*.

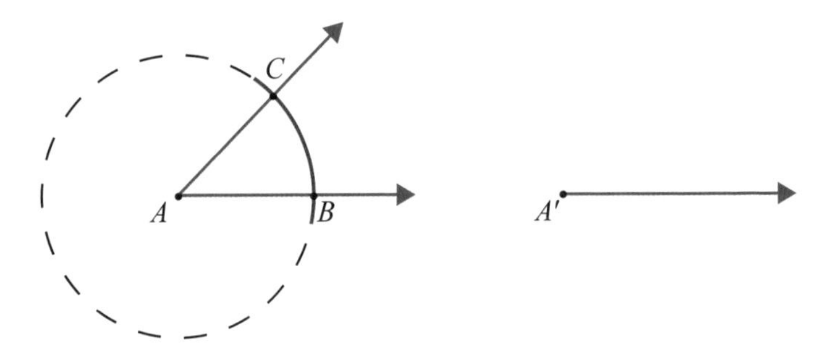

For this step, it's important to carefully hold the compass to the same radius as for the arc above. Position the needle point at *A*' and make a circle, or just a large arc that intersects the line segment. Label the intersection point *B*'.

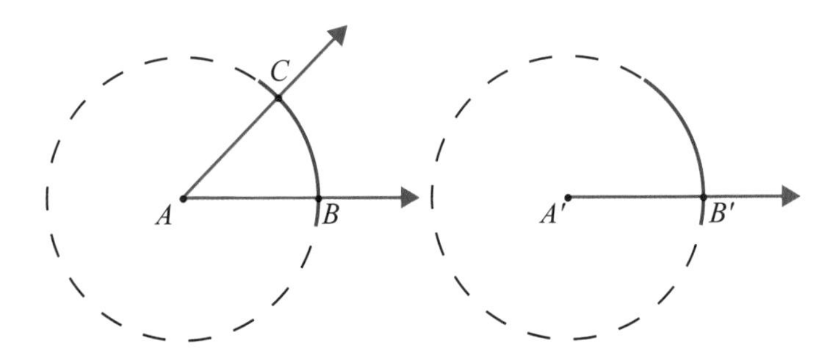

Next, use your compass to measure the distance between points *B* and *C*. Position the compass needle point at *B* and the drawing point at *C*. Carefully holding the compass, position the needle at point *B*', and draw an arc that intersects the other arc in your copy figure.

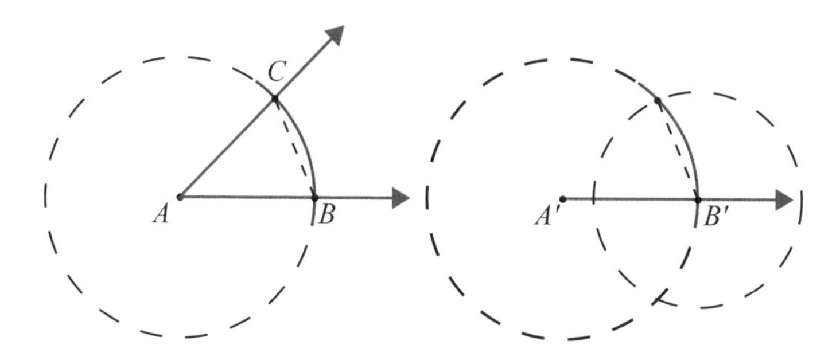

Finally, use your straightedge to draw a line through point *A*' and the intersection of the two arcs. This new, copied angle will be congruent to angle *A*.

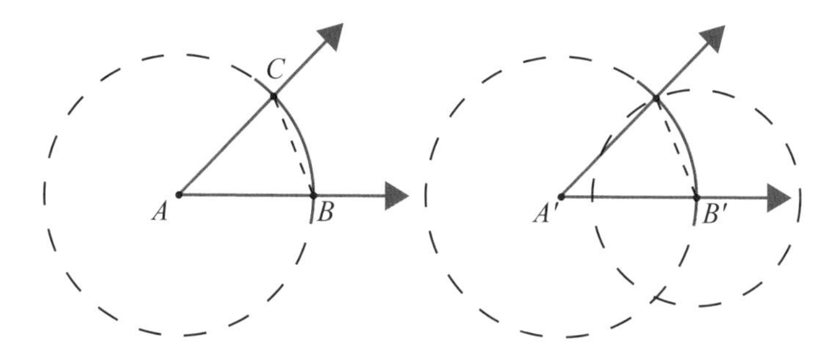

Try a few more for practice! Follow the steps above to construct an copy of an angle. Repeat until you're comfortable with the process.

PARALLEL LINE THROUGH A POINT

In this exercise, you'll construct a line that is parallel to another line, and also goes through a specified point. This method is based on our knowledge of parallel lines and transversals—that when two parallel lines are intersected by a transversal line, the alternate interior angles are congruent. Here, we'll copy an angle, using the strategy from the previous exercise, in order to create a parallel line.

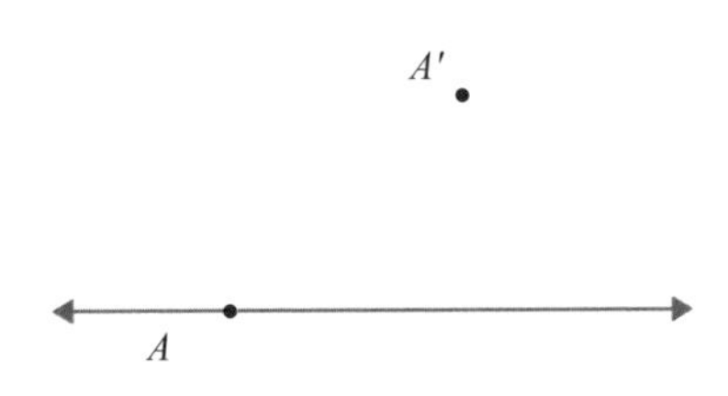

Begin with a line like the one above, with a point somewhere on the line (labeled *A*) and a point not on the line (labeled *A'*). Use your straightedge to draw a line segment that intersects *A* and *A'*. Make sure to extend the line a little beyond point *A'* so that you have space to copy the angle there later.

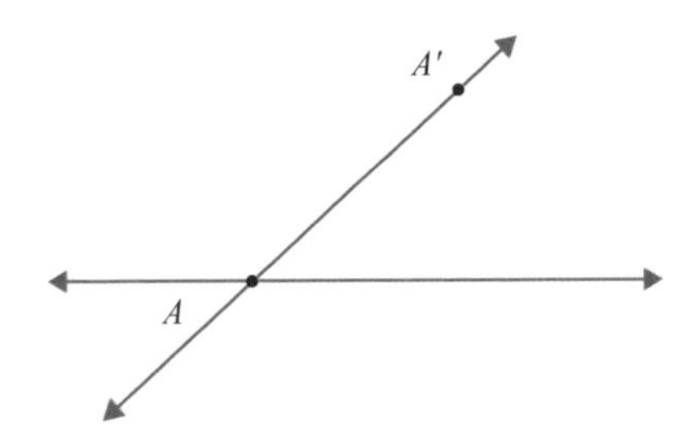

Next, using the strategy from the previous exercise, copy the angle formed by these two lines. Position the compass needle point on the vertex *A*, and make an arc that intersects the two line segments. Label the intersection points *B* and *C*.

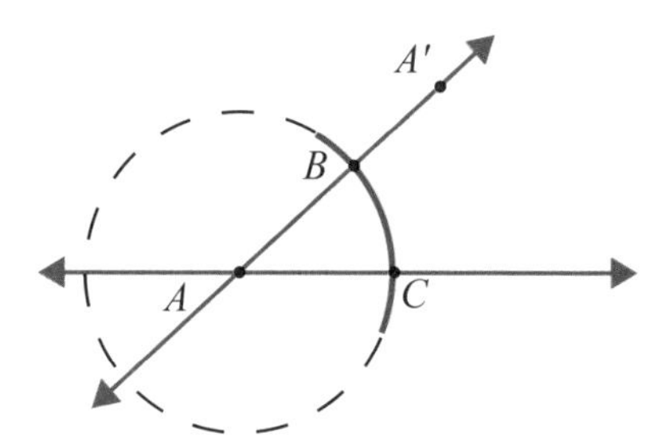

Carefully holding the compass to the same radius, position the needle at point *A*'. Draw an arc, and label the intersection point *B*'.

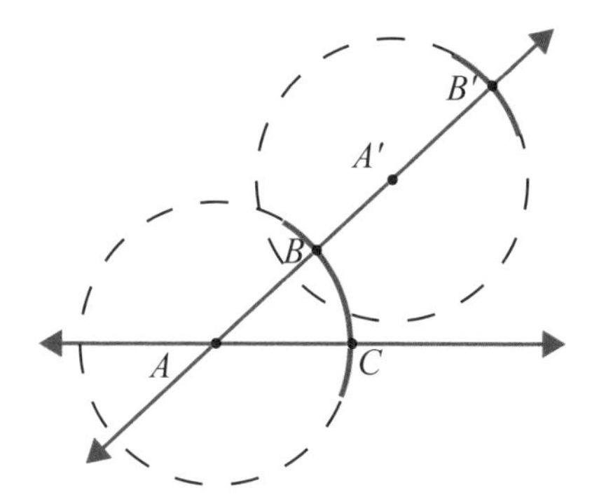

Next, use your compass to "measure" the distance between points *B* and *C*. Then, carefully holding the compass to the same radius, position the needle at point *B*', and draw an arc that intersects the other arc in your copy figure.

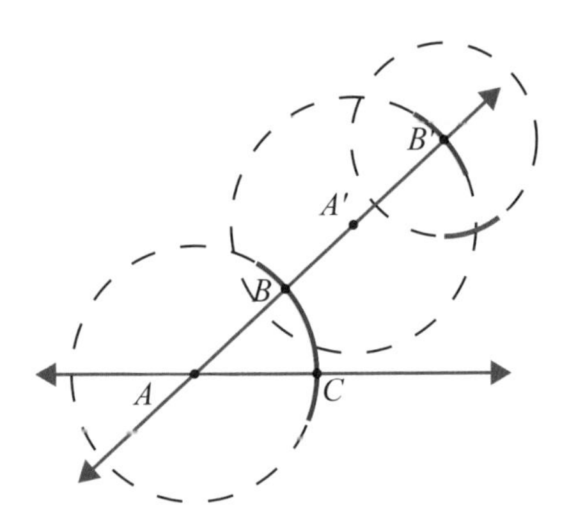

Finally, use your straightedge to draw a line through point *A*' and the intersection of the two arcs. This new, copied angle will be congruent to angle *A*. When a transversal creates congruent corresponding angles, the lines are parallel.

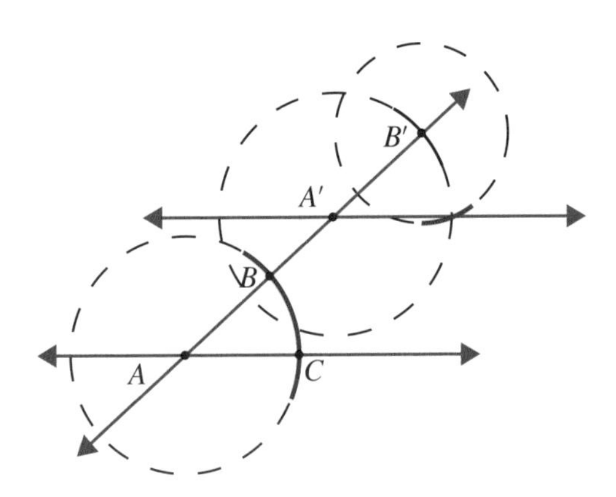

Try a few more for practice! Follow the steps above to construct a parallel line through a point. Repeat until you're comfortable with the process.

CONSTRUCTING POLYGONS

EQUILATERAL TRIANGLE

METHOD 1–START WITH TWO POINTS

In this exercise, you'll use your compass and straightedge to construct an **equilateral triangle.** The beginning steps are the same as the methods for creating **perpendicular lines**, seen earlier in this chapter. Therefore, we'll keep the instructions brief. Feel free to refer to the earlier section on constructing perpendicular lines for more details.

Supplies
For this section, you should have your compass and straightedge ready.

A • • B

Position the compass so that the needle is on point *A*, and the drawing point is on *B*. Make an arc above the two points in the spot where the third vertex of the triangle should go (eyeball it). You can also make a full circle if you prefer.

Constructing an equilateral triangle is very much like constructing perpendicular lines!

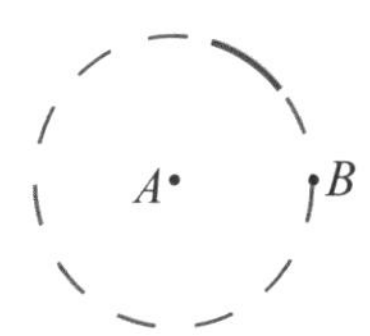

Next, position the compass so that the needle is on point *B*, and the drawing point is on *A*. With this radius, make a second arc to intersect the first one. Label this point *C*.

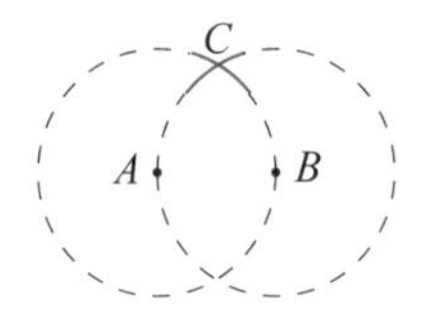

Finally, connect points *A*, *B*, and *C* to form a triangle. Use your straightedge.

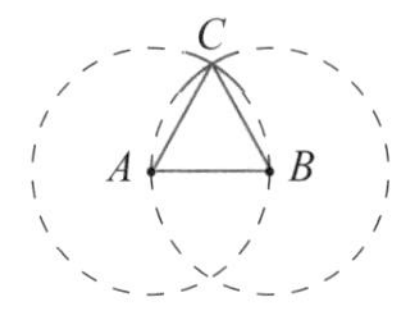

This triangle is equilateral. The reason this works is that the compass radius always matches the distance between *A* and *B*, so we know that all three sides are equal.

Try a few more for practice! Follow the steps above to create an equilateral triangle. Repeat until you're comfortable with the process.

METHOD 2–INSCRIBED IN A CIRCLE

For this method, begin with a circle, with a point marked for its center. Label the center point *A*. Also choose a point on the circle's circumference, and label it *B*.

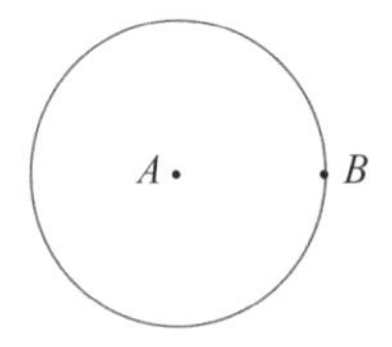

Position the compass so that the needle is on point *B*, and the drawing point is on *A*. With this radius, make two arcs that intersect the original circle. For the purposes of this exercise, label these intersection points *C* and *G*.

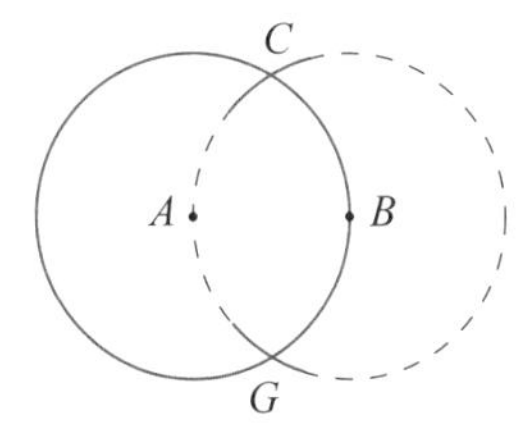

Next, position the compass so that the needle is on point *C*, and the drawing point is on *A*. With this radius, make an arc that intersects the original circle. Label this intersection point *D*.

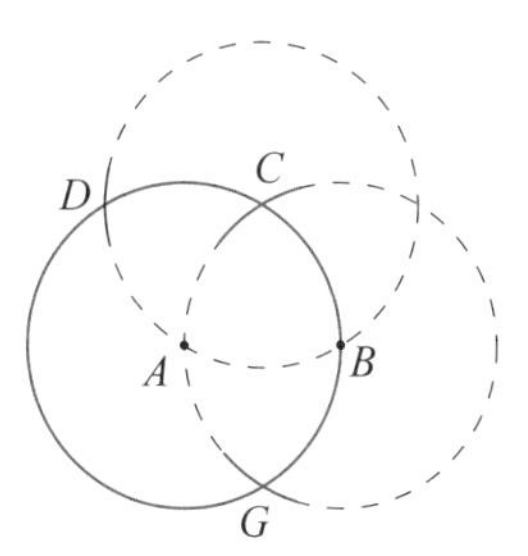

Next, position the compass so that the needle is on point *D*, and the drawing point is on *A*. With this radius, make an arc that intersects the original circle. Label this intersection point *E*.

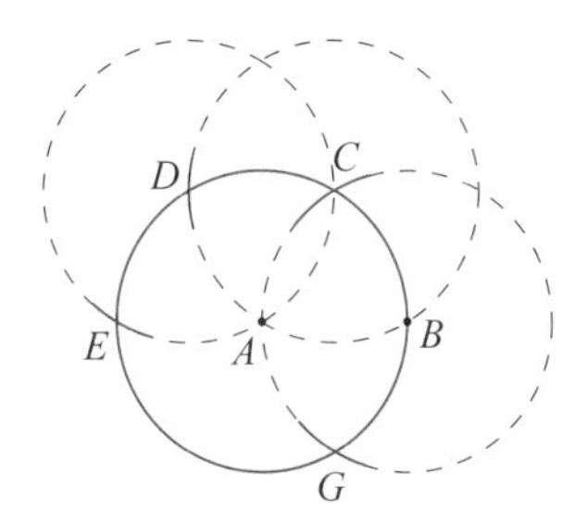

Next, position the compass so that the needle is on point *E*, and the drawing point is on *A*. With this radius, make an arc that intersects the original circle. Label this intersection point *F*.

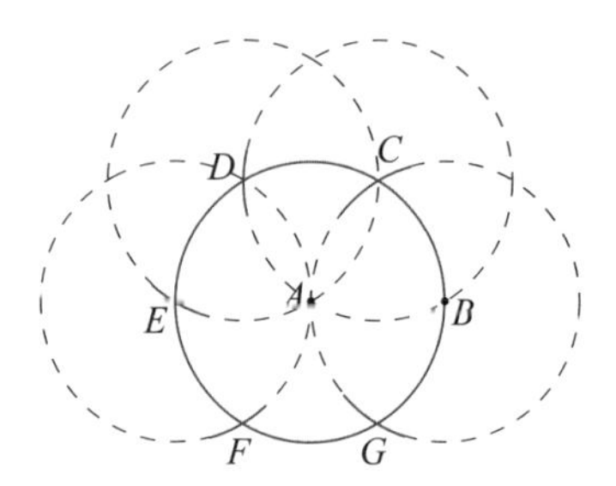

The final step is to connect points *B*, *D*, and *F* to make a triangle. (Alternatively, you could use points *C*, *E*, and *G*.)

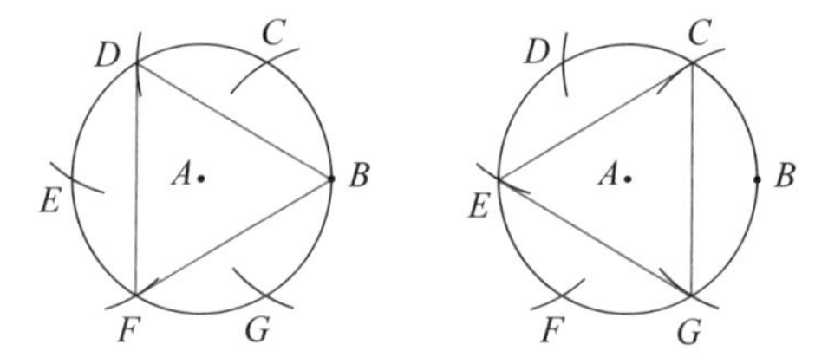

This triangle is equilateral, and it is **inscribed** in the circle. The reason this works is that we used the same compass radius for each arc we made—in other words, the intersection points were equally spaced from each other.

REGULAR HEXAGON

This method uses all the same steps as those used in the previous exercise (equilateral triangle inscribed in a circle). The difference is that at the end, you'll connect all six intersection points instead of just three.

Follow the steps in the previous exercise (equilateral triangle inscribed in a circle), proceeding until you've drawn all six arcs on the circle. See the figure below.

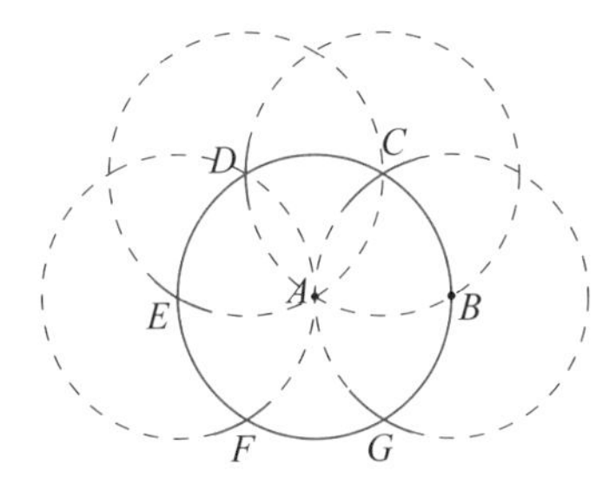

The final step is to connect these six intersection points to form a hexagon, as in the figure below. Use your straightedge.

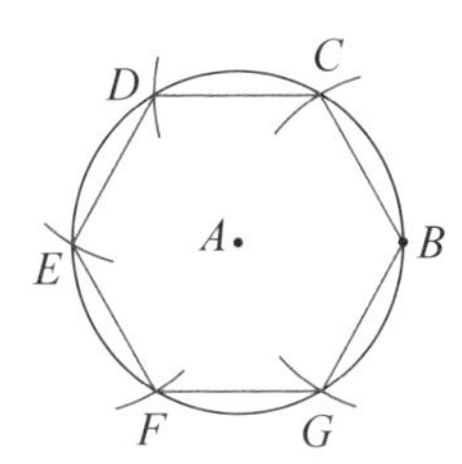

This hexagon is equilateral, and it is **inscribed** in the circle. The reason this works is that we used the same compass radius for each arc we made—in other words, the intersection points were equally spaced from each other.

SQUARE

METHOD 1–START WITH PERPENDICULAR LINES

In this exercise, you'll use your compass and straightedge to construct a **square.** For this method, start by following the steps for constructing **perpendicular lines**, seen earlier in this chapter. You can use either of the methods recommended. It might be a good idea to erase the circles as best you can, leaving only the perpendicular lines.

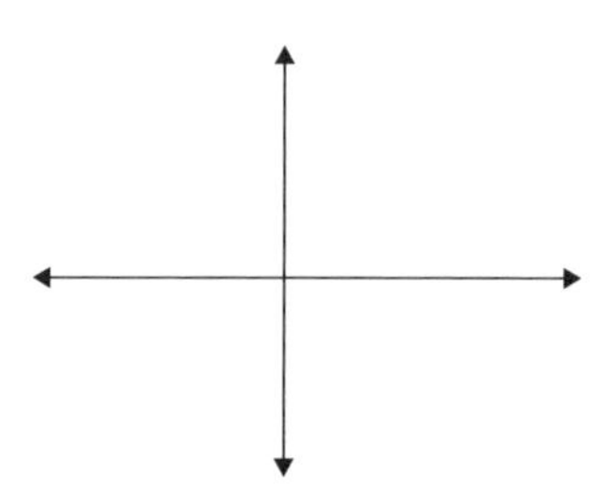

Once you have perpendicular lines, proceed to the next step. Position your compass needle at the intersection point of the two lines and make a circle. You can also just make arcs at the points where the circle crosses the two perpendicular lines.

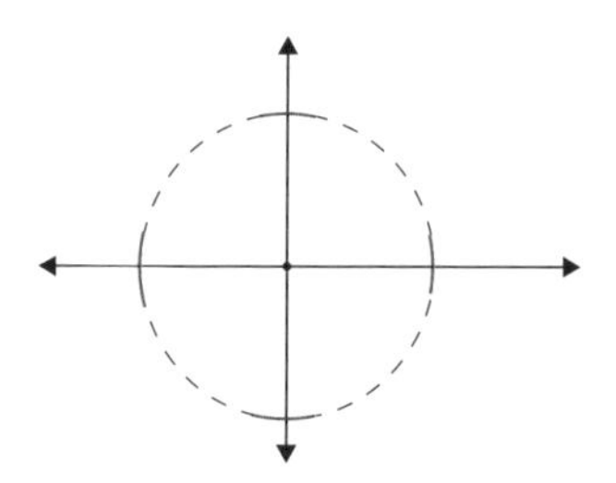

The circle crosses the two perpendicular lines, creating four intersection points. The next step is to connect those four intersection points to form a square, as in the figure below.

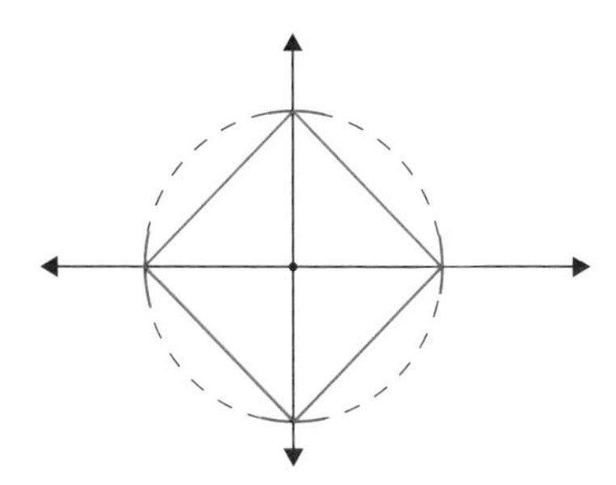

The quadrilateral drawn is a perfect square. One reason this works is that we know the two diagonals are perpendicular (we constructed them that way), and the four vertices are equidistant from the center.

Try a few more for practice! Follow the steps above to create a square. Repeat until you're comfortable with the process.

METHOD 2–INSCRIBED IN A CIRCLE

For this method, begin with a circle with a point marked for its center.

Position your straightedge carefully so that its line passes through the circle's center. Draw a straight line, which will be the diameter of the circle, or just mark the points where the straightedge intersects the circle. Label these points *A* and *B*.

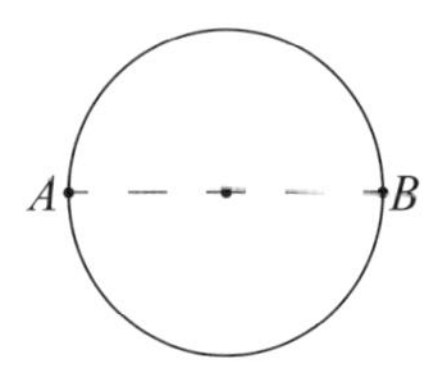

The next steps will be to construct a perpendicular bisector of this circle's diameter. Position the compass so that the needle is on point *A*, and the drawing point is on *B*. Make a circle, which will be larger than the given circle. Or, just make two arcs, one above and one below the diameter, spanning the spots where the perpendicular bisector will go (eyeball it).

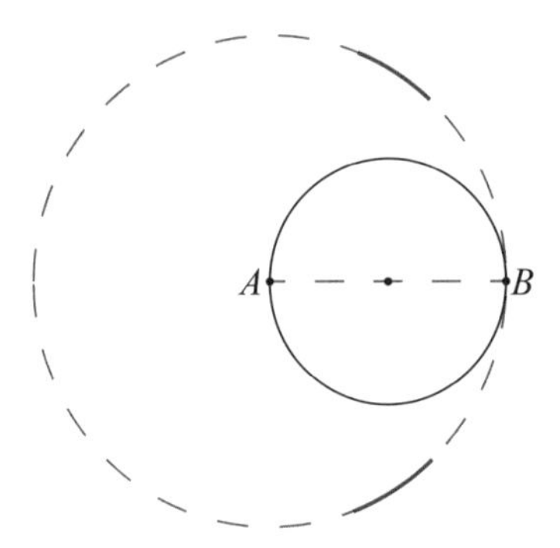

Next, position the compass so that the needle is on point *B*, and the drawing point is on *A*. Make a circle, or just two arcs, which intersects the previous arcs.

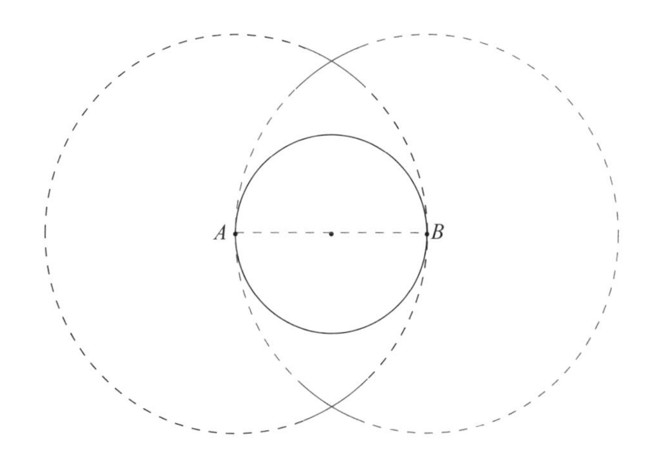

Position your straightedge to connect the two intersections of these larger arcs (it should also pass through the circle's center). Draw a straight line, or just mark the points where the straightedge intersects the original circle. Label these points *C* and *D*.

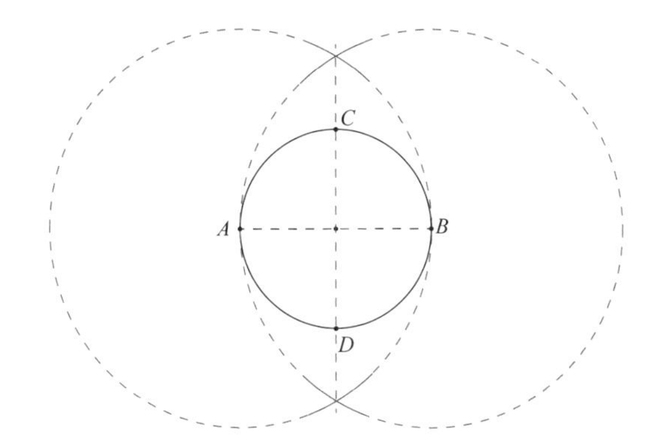

Finally, use your straightedge to connect points *A*, *B*, *C*, and *D*, forming a square.

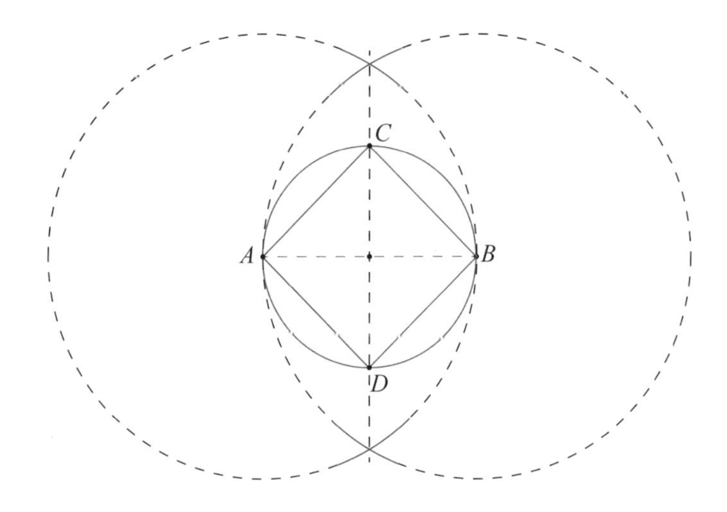

This square is **inscribed** in the circle.

Try a few more for practice! Follow the steps above to create a square. Repeat until you're comfortable with the process.

CIRCUMCENTER/CIRCUMCIRCLE

As we learned in Chapter 4, all triangles can have circumscribed circles (also known as **circumcircles**). The **circumcenter** of a triangle is the intersection of the **perpendicular bisectors** of its sides.

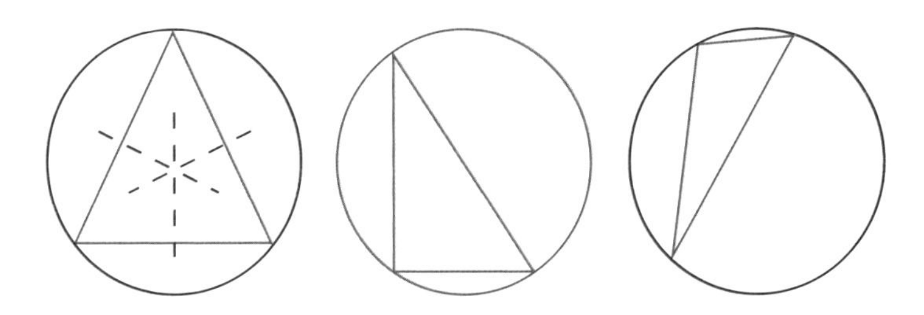

To find the circumcenter, you'll construct the perpendicular bisectors and see where they meet.

When three or more lines intersect, the intersection point is known as a **point of concurrency**.

Then, to construct the circumcircle, you'll center the circle at the constructed circumcenter, and make the circle touch the triangle's vertices.

Here's how to do it.

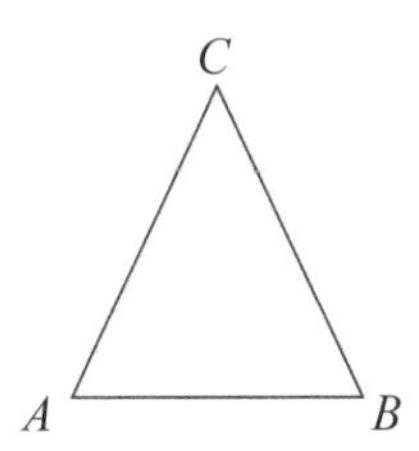

We will construct the circumcircle of triangle *ABC*. To find the circumcenter of the triangle, construct the triangle's perpendicular bisectors. It will be sufficient to construct just two of the three bisectors, since that will be enough to show the point of concurrency.

1. Construct the perpendicular bisector for side *AB*.

With the compass needle on vertex *A* and the drawing point on *B*, make a circle.

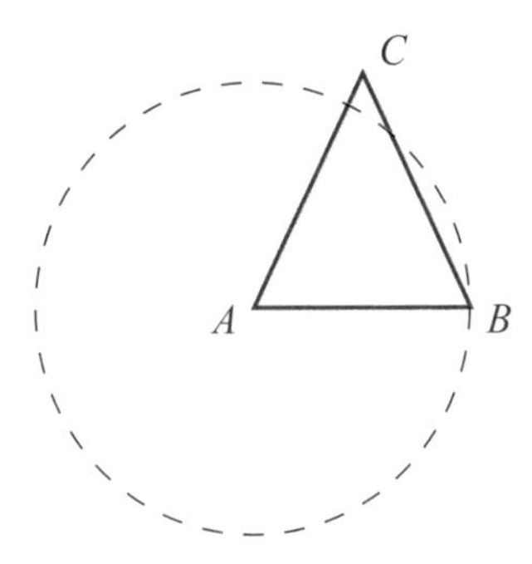

Then, with the compass needle on *B* and the drawing point on *A*, make another circle, which will intersect the first circle in two places.

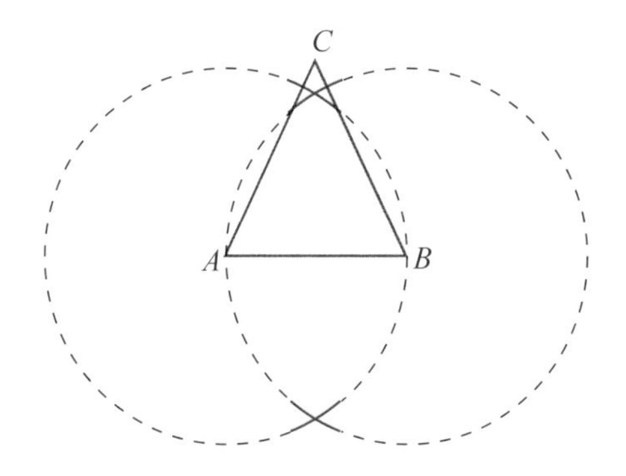

Next, use a straightedge to connect the two points where the arcs intersect, as shown. This new line is the perpendicular bisector for *AB*.

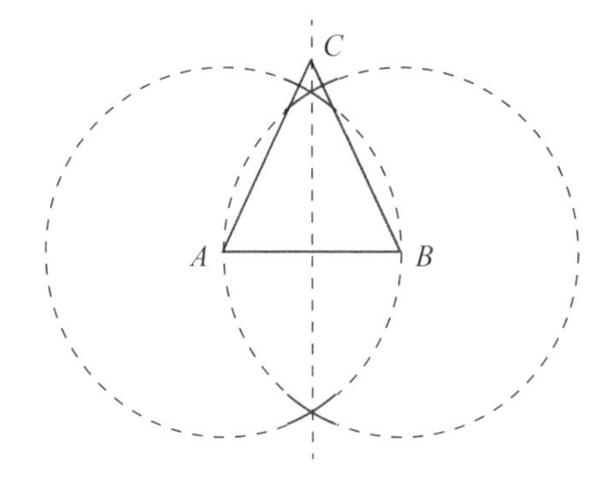

2. Construct the perpendicular bisector for side *BC*.

Repeat the same process to construct the perpendicular bisector for side *BC*. You may find it helpful to erase some of the previous arcs so that you can better see what you're doing.

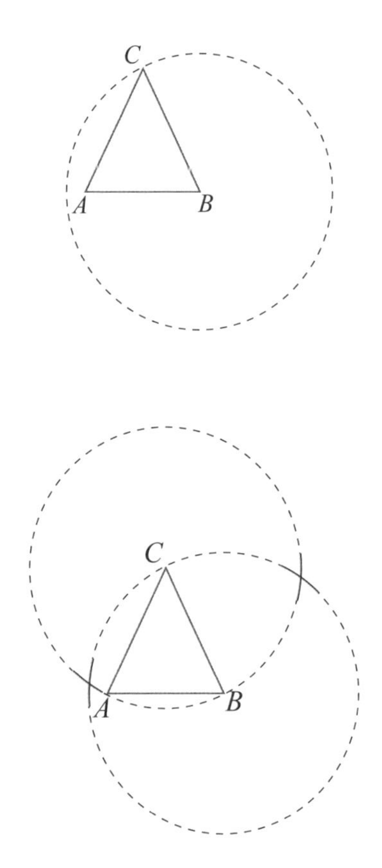

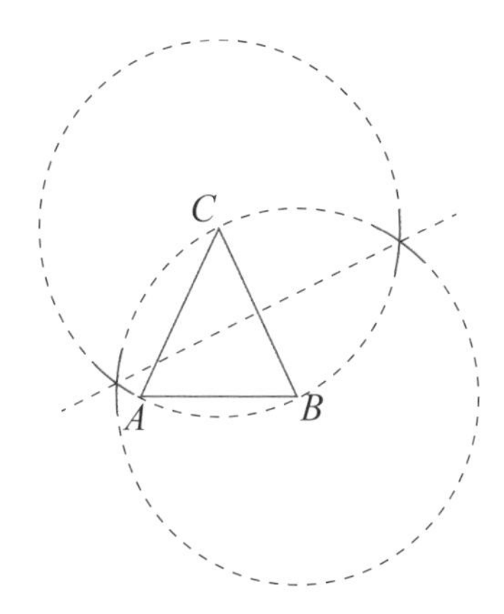

We have found the circumcenter of triangle *ABC*. For this exercise, we'll label the circumcenter as *O*.

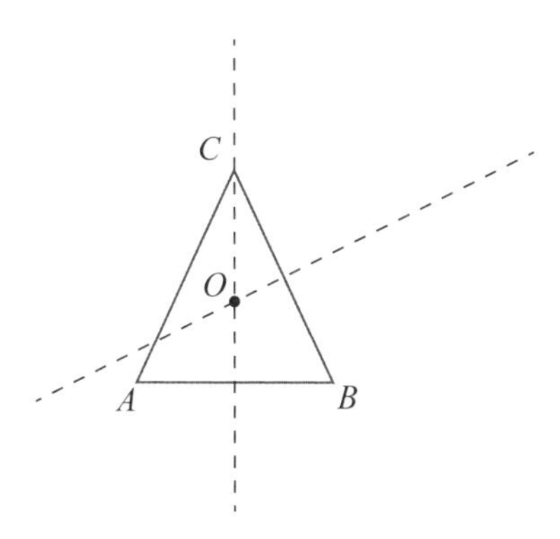

3. Draw the circumcircle.

Next, make a circle that touches the vertices of the triangle. Place the compass needle on point O and the drawing point on one of the vertices. Make a circle. If you did it correctly, you'll find that the circle touches each of the triangle's vertices.

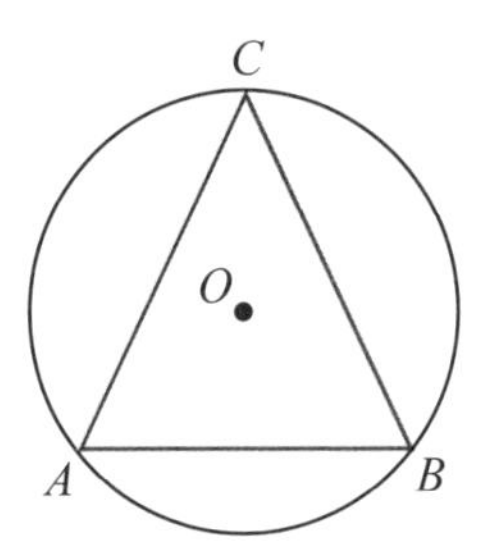

This method works to construct a circumcircle for *any* triangle. Try it on the ones below! Then, try a few more on your own for practice.

SPECIAL CASE: RIGHT TRIANGLE

In a right triangle, the circumcenter will always be the midpoint of the hypotenuse. You can construct the circumcenter more easily by bisecting the hypotenuse.

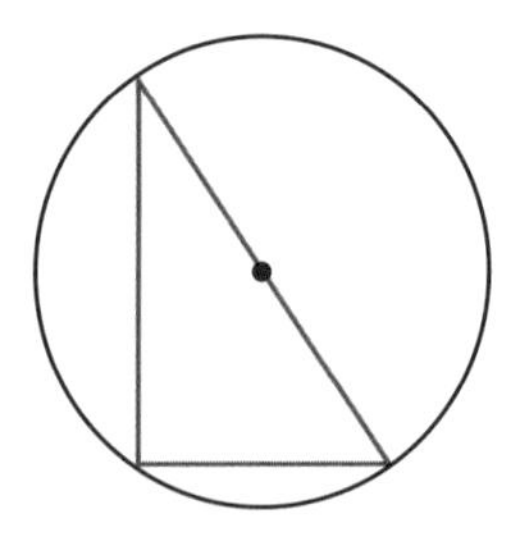

INCENTER/INCIRCLE

All triangles can have inscribed circles (also known as **incircles**). The **incenter** of a triangle is the point where the **angle bisectors** intersect. To find the incenter, you'll construct the angle bisectors of the triangle and see where they meet.

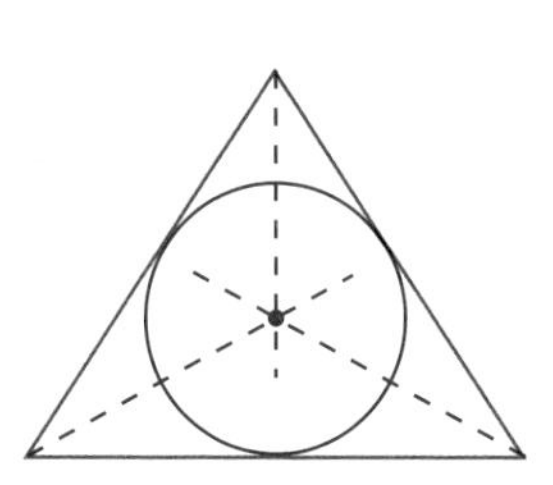

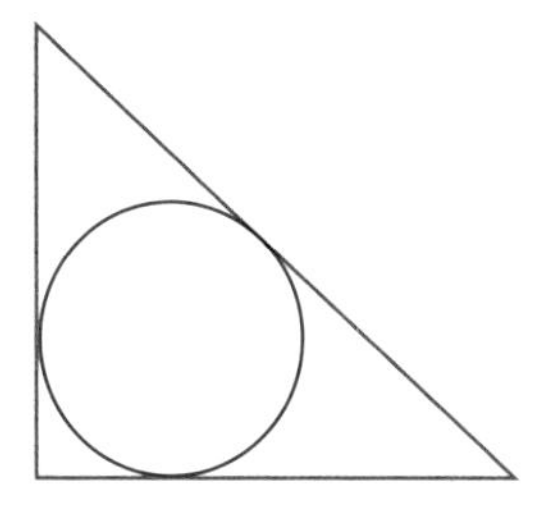

The incircle will be **tangent** to the triangle's sides. To find one of these points of tangency, you'll construct a perpendicular line that passes through the incenter.

Here's how to do it.

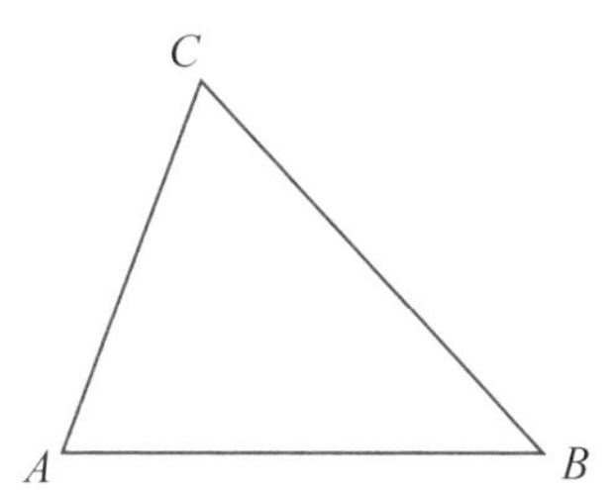

Construct the inscribed circle of triangle *ABC*.

To find the incenter of the triangle, construct the triangle's angle bisectors. We need to construct only two of the three angle bisectors, since that will be enough to show the point of concurrency.

1. Construct the angle bisector for angle *A*.

With the compass needle on vertex *A*, make an arc that passes through the two legs of that angle. For the purposes of this exercise, we'll call these two intersection points *M* and *N*.

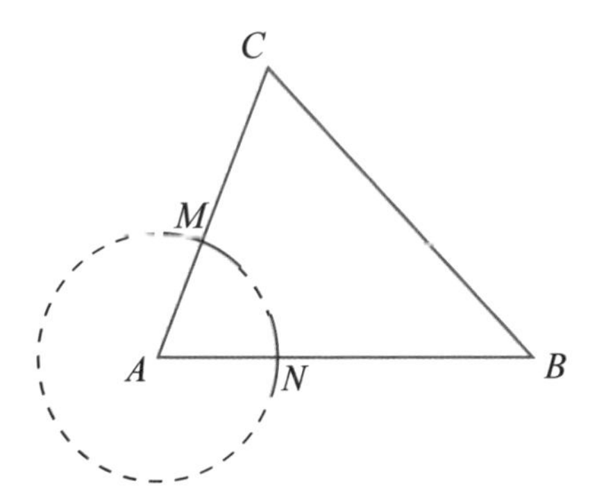

Now, position the compass needle point on *M* and the drawing point on *A*. With this radius, make a circle.

Repeat for point *N*, making a circle of the same radius, which will intersect the previous circle.

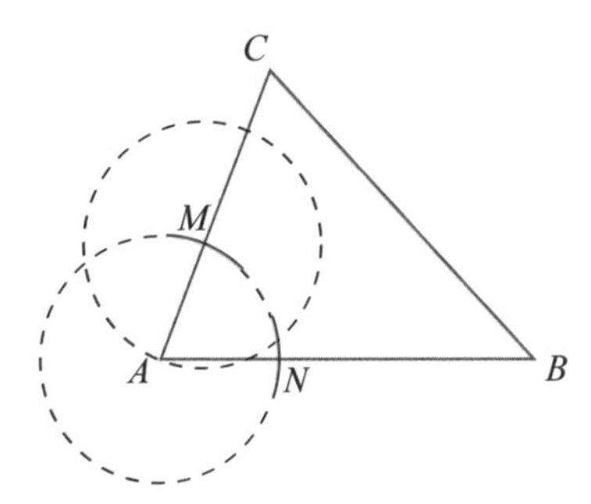

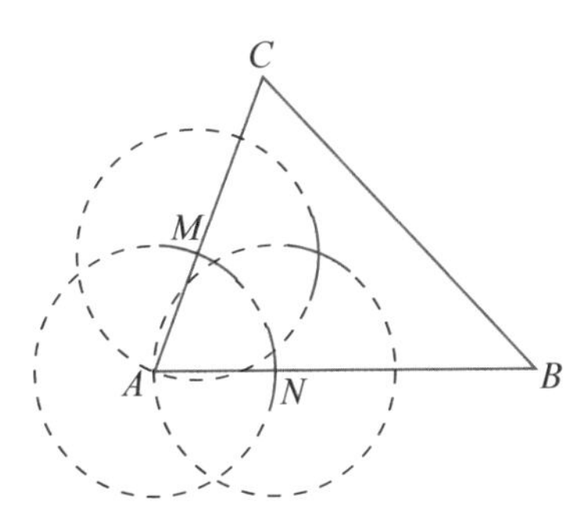

Finally, use a straightedge to draw a line through this new intersection and point *A*.

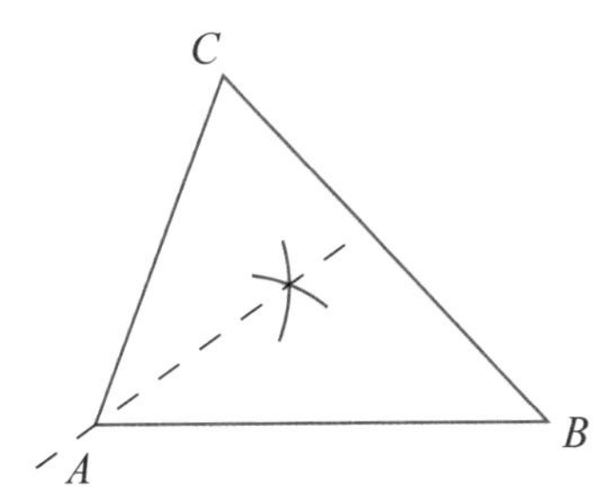

2. Construct the angle bisector for angle *B*.

Repeat the same process to construct the angle bisector for angle *B*. You may find it helpful to erase some of the previous arcs so that you can better see what you're doing.

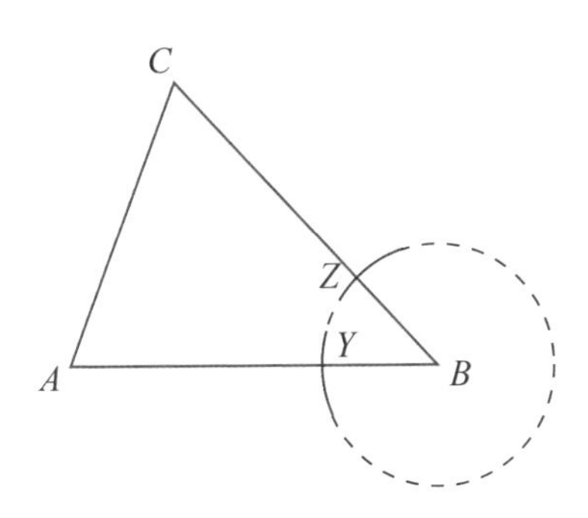

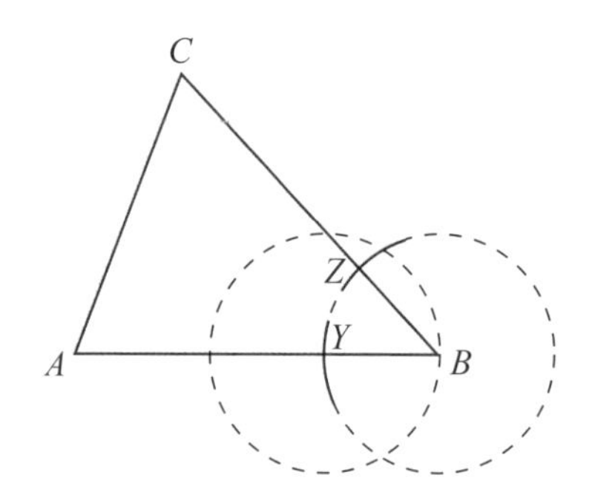
C
Z
Y
A
B

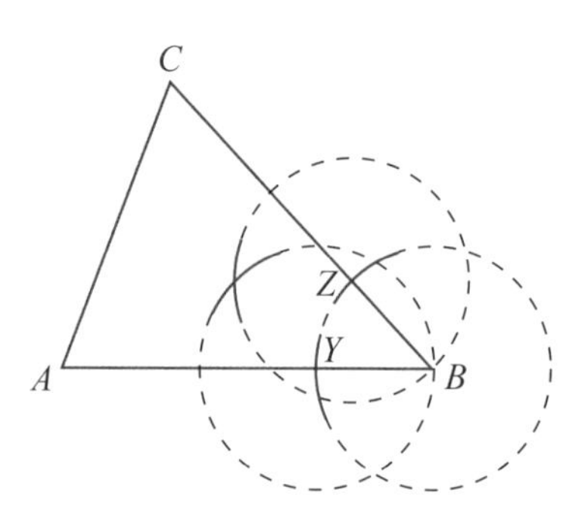
C
Z
Y
A
B

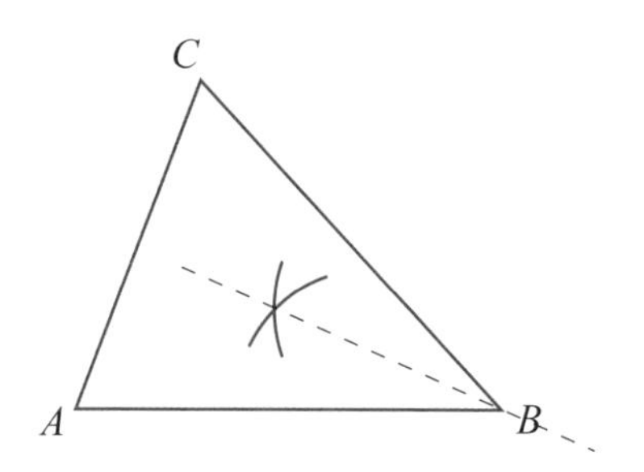
C
A
B

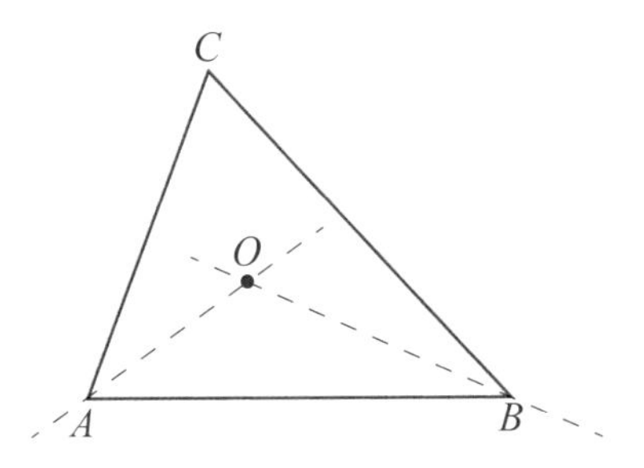

We have found the incenter of triangle *ABC*. For this exercise, we'll label the incenter as *O*.

Next, make a circle that is tangent to the sides of the triangle. In order to make sure that it's tangent, construct a line perpendicular to one of the sides. You need to do this only; this will correctly identify the radius of our circle.

3. Construct a perpendicular line through the incenter.

With the compass needle on the incenter *O*, make an arc that intersects side *AB* in two places. For this exercise, we'll call these intersection points *P* and *Q*.

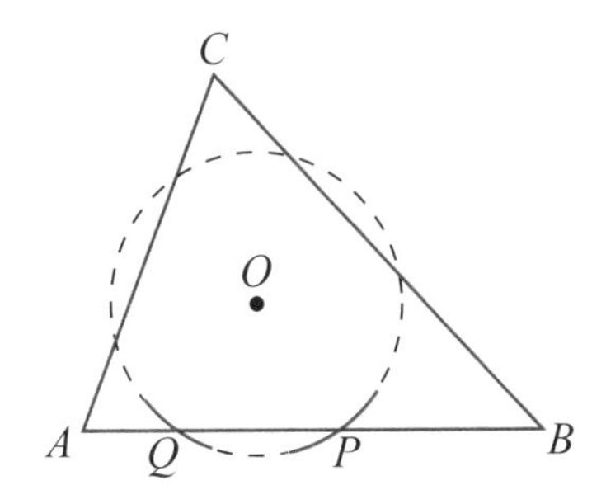

Now, with the compass needle point on P and the drawing point on O, make a circle.

Repeat for point Q, making a circle of the same radius, which will intersect the previous circle.

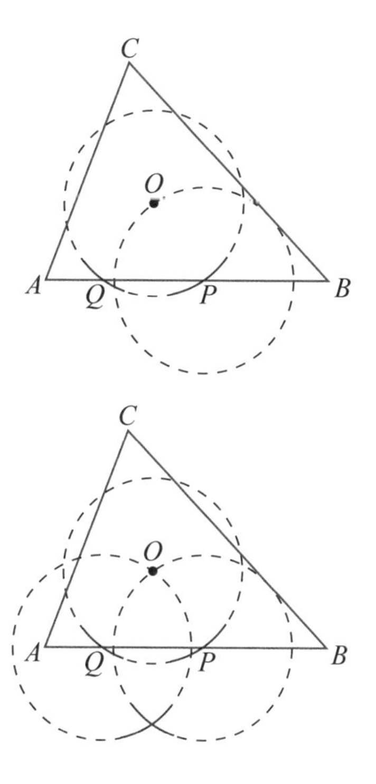

Use the straightedge to draw a line through this new intersection and point O.

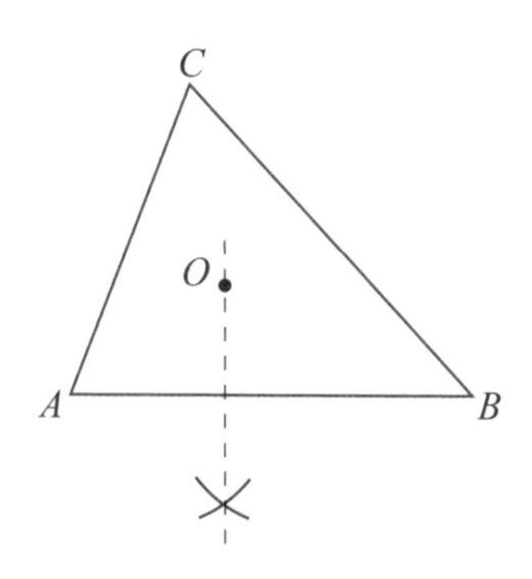

4. Draw the incircle.

This perpendicular intersection between the new line and side *AB* is one of the three tangent points for our circle. Now, the circle's center and radius are defined, and the circle can be drawn.

Position the compass needle at point *O*, and the drawing point at the perpendicular intersection that you have just drawn. Finally, make a circle. If you did it correctly, you'll find that the circle is tangent to each of the three sides of the triangle.

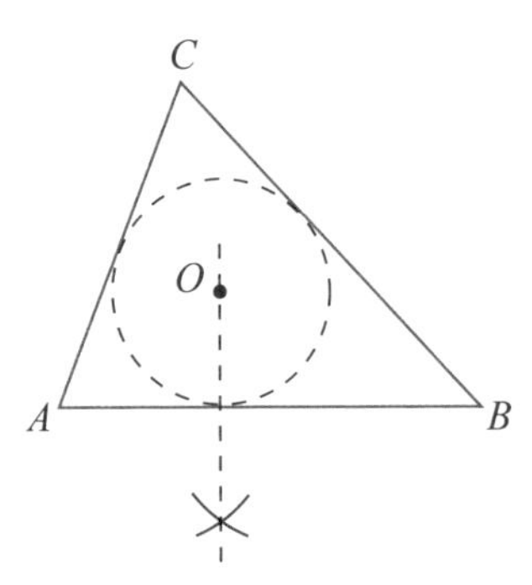

This method works to construct an incircle for *any* triangle. Try it on the ones below! Then, try a few more on your own for practice.

THE CENTROID AND ORTHOCENTER

CENTROID

The centroid is the intersection of the **medians** of a triangle. A median is a segment from the triangle's vertex to the midpoint of the opposite side.

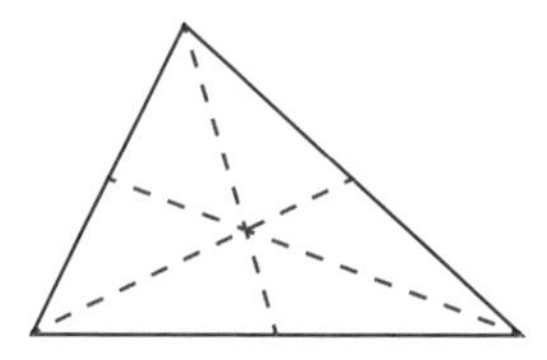
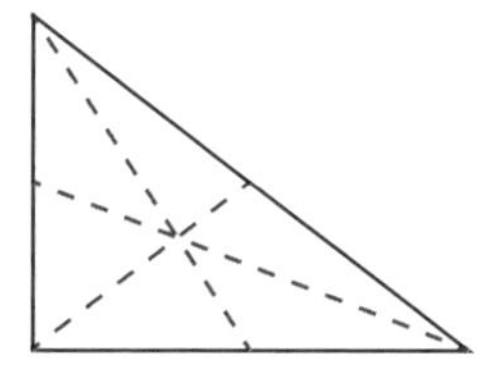

To construct a median of a triangle, first bisect one of the triangle's sides. Then, connect the midpoint to the opposite vertex. Construct the remaining two medians to find the centroid.

Try it on the triangles below!

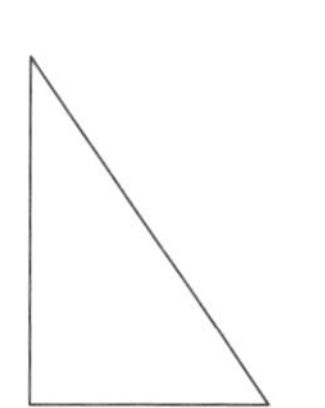

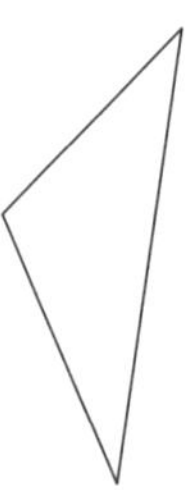

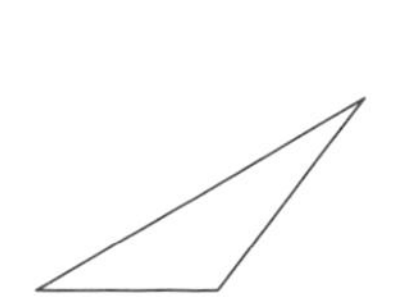

ORTHOCENTER

The orthocenter is the intersection of the **altitudes** of a triangle. An altitude is a segment from the triangle's vertex that is perpendicular to the opposite side. The orthocenter can be outside the triangle.

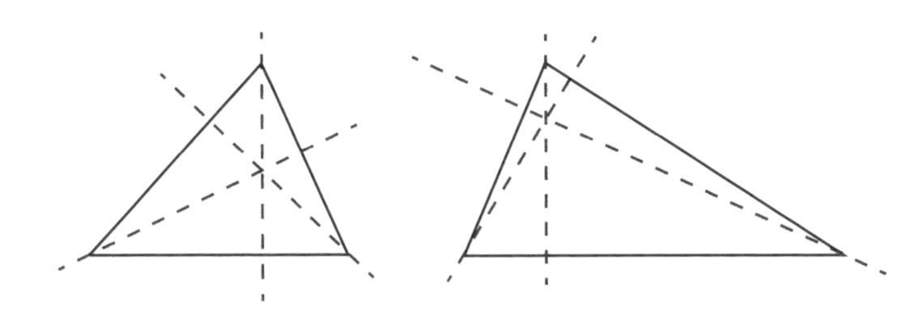

To construct an altitude of a triangle, construct a perpendicular line that passes through the opposite vertex. Construct two altitudes to find the orthocenter.

Try it on the triangles below!

SPECIAL CASE: EQUILATERAL TRIANGLE

In an equilateral triangle, the incenter, circumcenter, centroid, and orthocenter are *all* found at the same point!

CIRCUMCENTER OF A QUADRILATERAL

As mentioned in Chapter 4, not all quadrilaterals have circumcircles. In fact, one of Euclid's discoveries was that a quadrilateral will have a circumcircle *only* if its opposite angles are **supplementary**. You won't always know the angles of a quadrilateral, so in order to construct a [possible] circumcircle, you may just go through a little trial and error.

If a quadrilateral (or other polygon) does have a circumcenter, then its definition is the same as that for a triangle—the circumcenter is the intersection of the perpendicular bisectors of the polygon's sides. To attempt construction of a polygon's circumcircle, construct each of the perpendicular bisectors of its sides. If and only if the perpendicular bisectors intersect at a single point, then the circumcircle can be constructed.

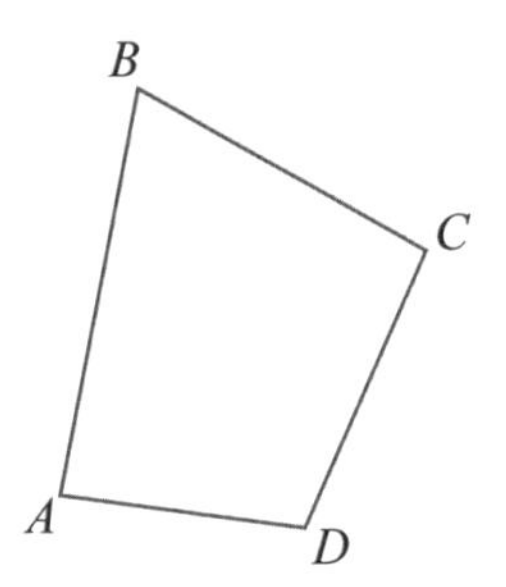

Construct the circumcircle for quadrilateral *ABCD*.

If you know that a quadrilateral has a circumcenter, then you need to construct only two perpendicular bisectors in order to find the circumcenter. It's more common that you wouldn't know whether or not the circumcenter can be constructed, so make a habit of constructing all four perpendicular bisectors.

First, we can tell you that this quadrilateral does, in fact, have a circumcenter. To find it, construct the perpendicular bisectors of the quadrilateral's sides, and see where they intersect. For this exercise, we will assume that you are comfortable with constructing perpendicular bisectors.

1. Construct the perpendicular bisectors for each side.

Construct the perpendicular bisector for AB:

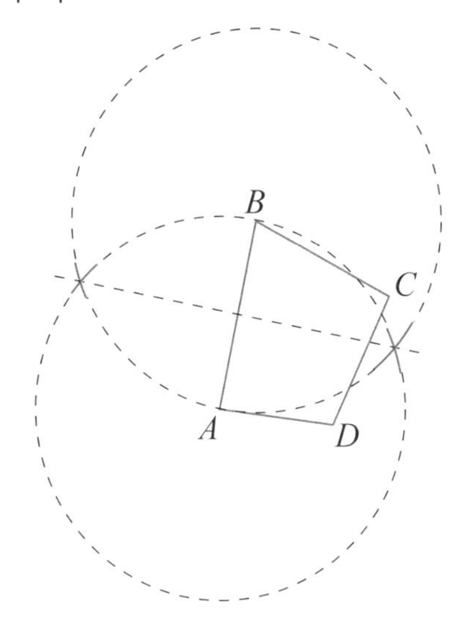

Construct the perpendicular bisector for BC:

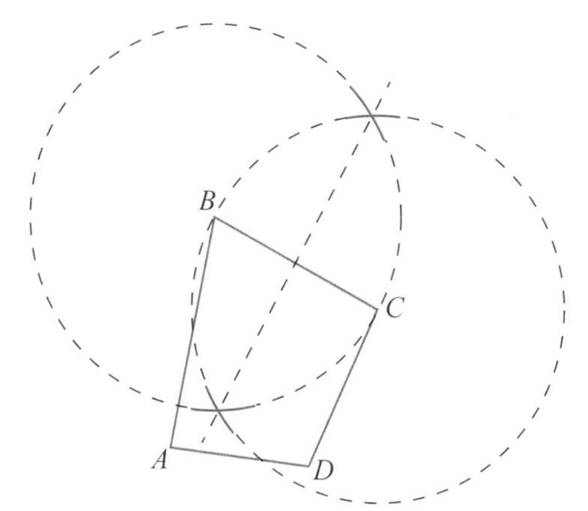

Construct the perpendicular bisector for CD:

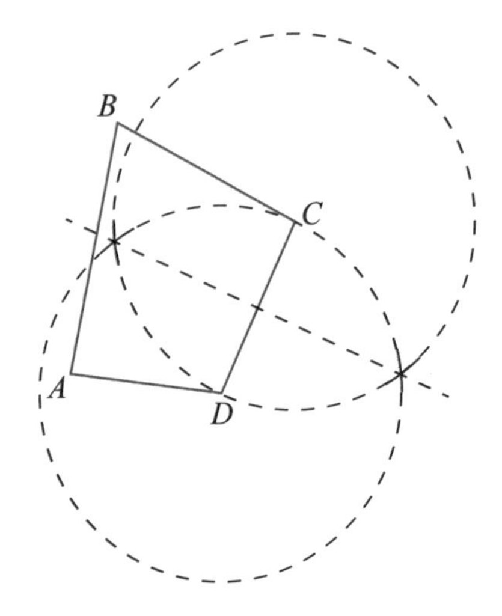

Construct the perpendicular bisector for DA:

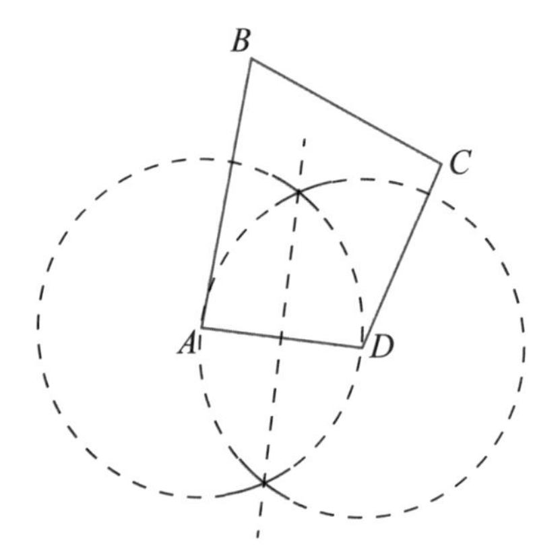

We have found the circumcenter of the quadrilateral.

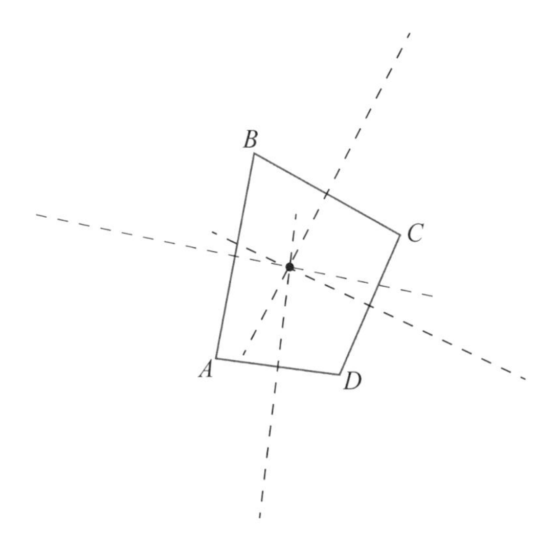

2. Draw the circumcircle.

Finally, construct the circle centered at the circumcenter and touching the vertices of the quadrilateral.

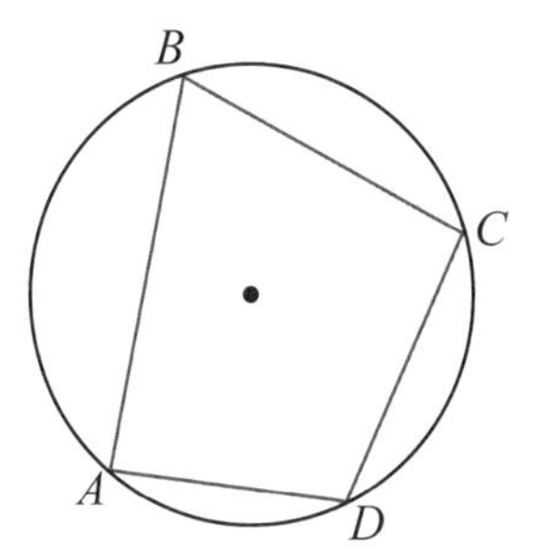

INCENTER OF A QUADRILATERAL

As mentioned in Chapter 4, not all quadrilaterals have incircles. One property of quadrilaterals that have incircles is that the two pairs of opposite sides have the same total length. You won't always know the side lengths of a quadrilateral, so in order to construct a [possible] incircle, you may have to just go through a little trial and error.

If a quadrilateral (or other polygon) does have an incircle, then its definition is the same as that for a triangle—the incenter is the intersection of the angle bisectors of the polygon. To attempt construction of a polygon's incircle, construct each of the angle bisectors. If and only if the angle bisectors intersect at a single point, then the incircle can be constructed.

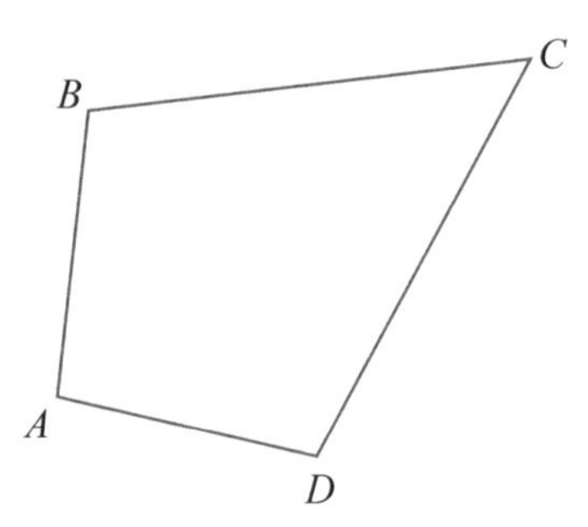

Construct the incircle for quadrilateral *ABCD*.

First, we can tell you that this quadrilateral does, in fact, have an incenter. To find it, construct the angle bisectors of the quadrilateral, and see where they intersect. For this exercise, we will assume that you are comfortable with constructing angle bisectors.

1. Construct the angle bisectors for each angle.

Construct the angle bisector for *A*:

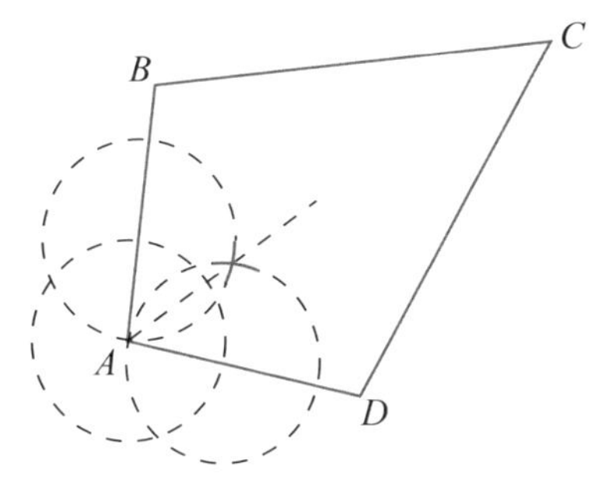

Construct the angle bisector for *B*:

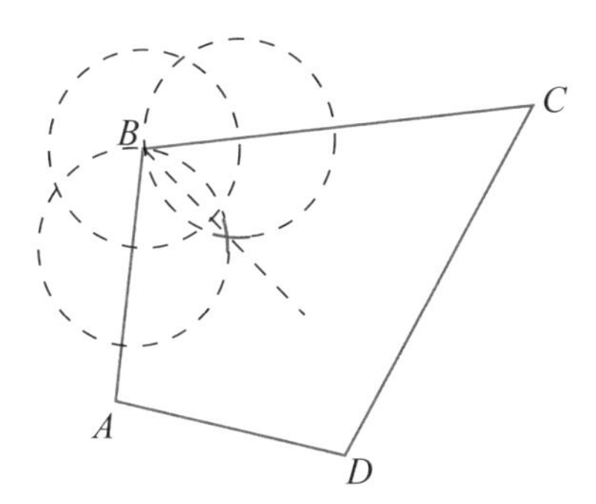

Construct the angle bisector for *C*:

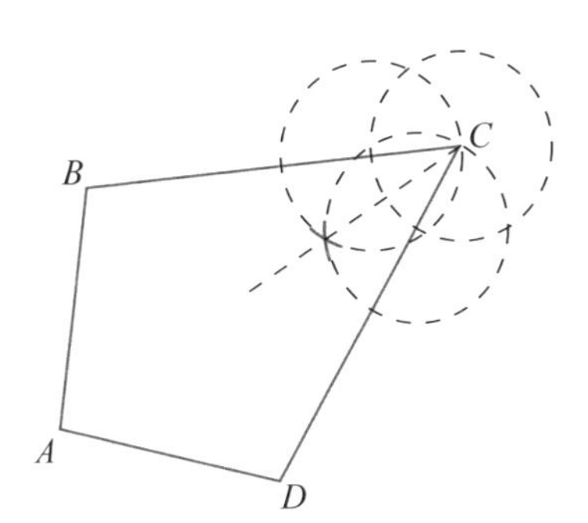

Construct the angle bisector for *D*:

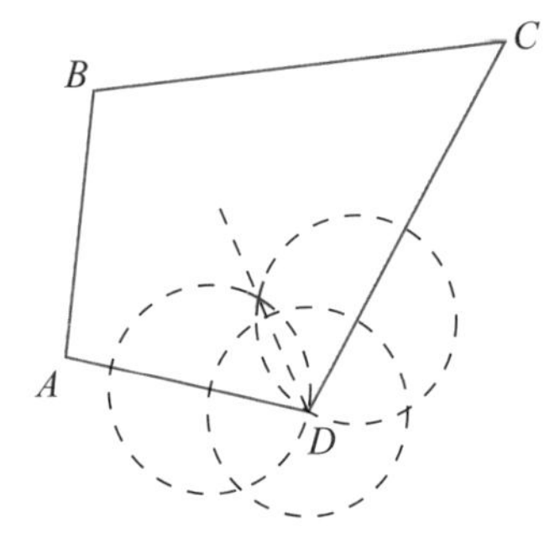

We have found the incenter of quadrilateral *ABCD*. For this exercise, we'll label the incenter as *O*.

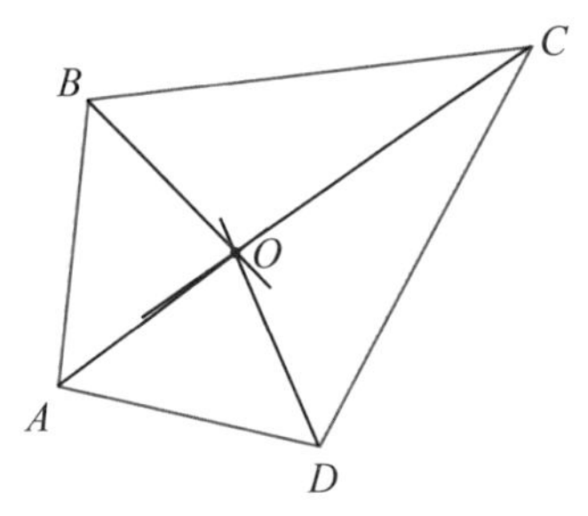

Next, make a circle that is tangent to the sides of the quadrilateral. In order to make sure that it's tangent, construct a perpendicular line to one of the sides. You need to do this only once; this will correctly identify the radius of the circle.

2. Construct a perpendicular line through the incenter.

With the compass needle on the incenter *O*, make an arc that intersects side *AB* in two places. For this exercise, we'll call these intersection points *P* and *Q*.

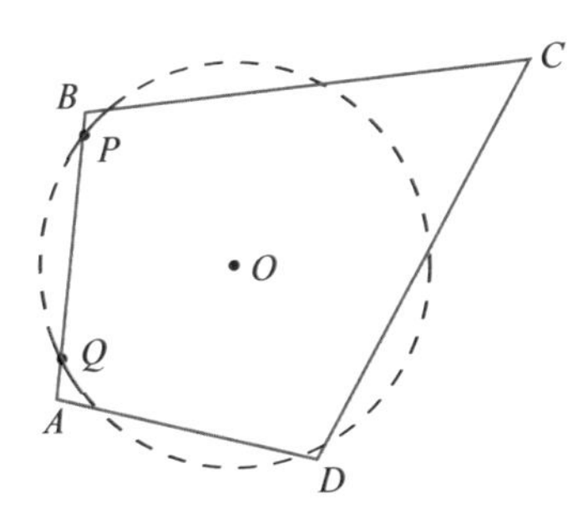

Now, position the compass needle point on *P* and the drawing point on *O*. With this radius, sweep the compass around and make an arc that's outside of the quadrilateral, on the opposite side of *AB*.

Repeat for point *Q*, making an arc of the same radius, which will intersect the previous arc.

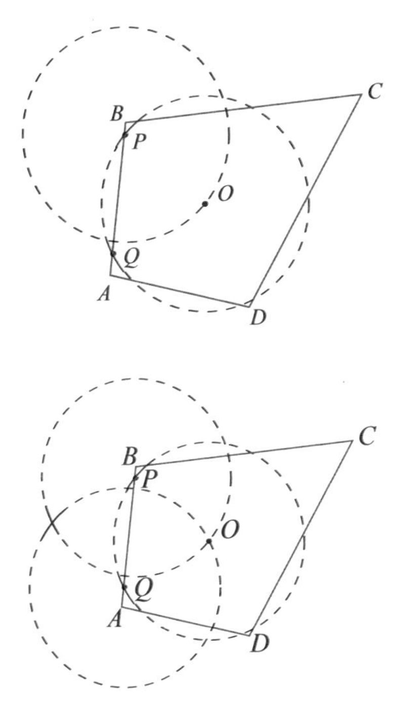

Use the straightedge to draw a line through this new intersection and point O.

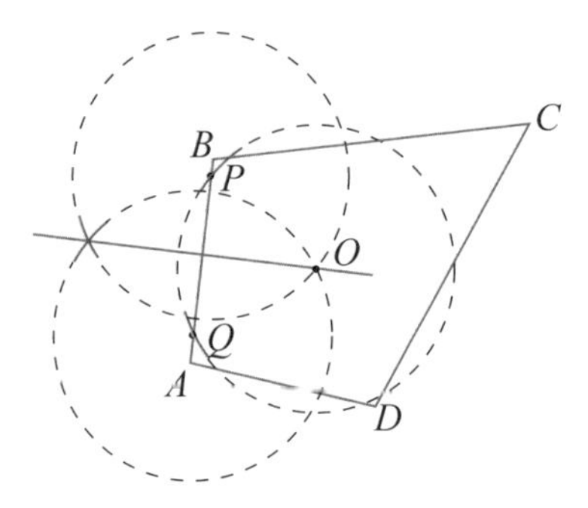

3. Draw the incircle.

This perpendicular intersection between the new line and side AB is one of the four tangent points for our circle. Now, the circle's center and radius are defined and the circle can be drawn.

Position the compass needle at point O, and the drawing point at the perpendicular intersection that you have just drawn. Finally, make a circle. If you did it correctly, you'll find that the circle is tangent to each of the four sides of the quadrilateral.

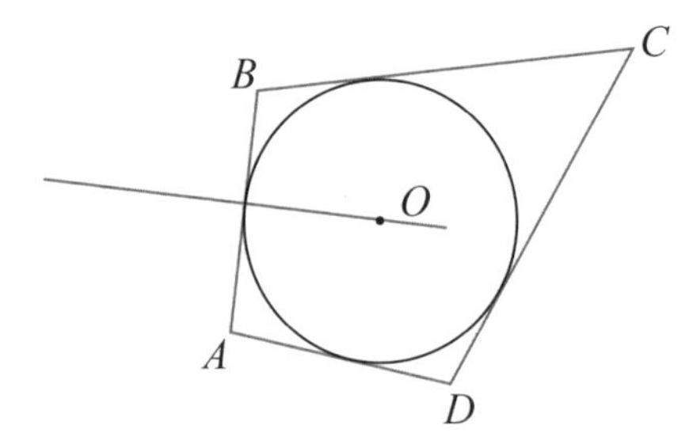

CHAPTER

CONNECTING ALGEBRA AND GEOMETRY

© istockphoto.com / Bilanol

Algebra and Geometry are often taught as two separate courses. However, these are actually closely related concepts. Algebra can often be used to help express ideas and solve problems in geometry that would be a lot more difficult otherwise.

CHAPTER CONTENTS

ALGEBRA AND THE COORDINATE PLANE

COORDINATE GEOMETRY BASICS

EQUATION OF A LINE

Equation of a Line

$$y = mx + b$$

An equation in this form is called a **linear equation**, because it is guaranteed to form a straight line when graphed. The following variables make up this equation:

- x and y are the coordinates (x, y) of a point on the line.
- m is the slope of the line (rise/run).
- b is the y-intercept of the line (the point where $x = 0$).

In most cases, when dealing with the equation of a line, you'll see x and y as variables, and numerical values for m and b. Such an equation represents all the possible (x, y) coordinates that exist on a specific line. Consider this example:

$$y = 3x + 4$$

In this example, any (x, y) coordinate pair that exists on the line will satisfy the equation, and any coordinate pair that satisfies the equation will exist on the line. The equation is a function—if you plug in a value for x, you'll get a corresponding value for y. For instance, if we plug in $x = 2$, then $y = 10$:

$$y = 3(2) + 4$$
$$y = 6 + 4$$
$$y = 10$$

Therefore, we know that the coordinate $(2, 10)$ exists on the line $y = 3x + 4$.

In coordinate geometry, you will often need to solve for one of the variables in a linear equation. For example, you might solve for slope (m) in order to find out if two lines are parallel. Or, you might solve for the x- or y-coordinate so that you can plot a point on a line.

For example, suppose the coordinate $(2, 5)$ lies on the line with the equation $y = 3x + b$. What is the value of b?

In this example, there is only one unknown variable, b. If you plug in $x = 2$ and $y = 5$ (from the given coordinate), you can solve:

$$y = 3x + b$$
$$5 = 3(2) + b$$
$$5 = 6 + b$$
$$5 - 6 = b$$
$$\mathbf{-1 = b}$$

FIND A LINE FROM AN EQUATION

To graph a line, you just need to plot two points, and then connect them. You can find points that lie on a line by choosing different values for x and then evaluating for y, or vice versa.

In this example, we will graph the line represented by the equation $y = 2x + 5$. Begin by finding a coordinate pair that satisfies the equation. To find a coordinate pair, choose a value for x and then evaluate for y. Any value will do! Try to choose numbers that make the math easy.

If you plug in $x = 4$, then you get $y = 13$.

$$y = 2x + 5$$
$$y = 2(4) + 5$$
$$y = 8 + 5$$
$$y = 13$$

Therefore, the coordinate (4, 13) lies on the line.

Now, you just need one more coordinate pair to make a line. Again, any value will do.
Try $x = -6$.

$$y = 2x + 5$$
$$y = 2(-6) + 5$$
$$y = -12 + 5$$
$$y = -7$$

Therefore, the coordinate (−6, −7) lies on the line.

Finally, plot the two points, and connect them to make a line.

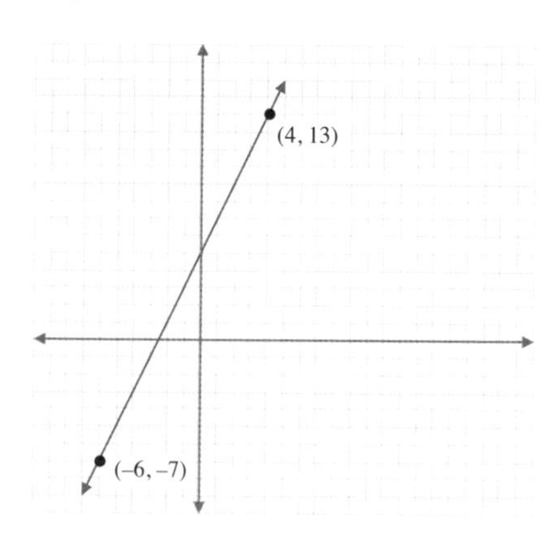

FIND AN EQUATION FROM A LINE

To derive an equation from a line, you can start by choosing any two points from the line. With two points, you can then solve for *m* and *b*.

If the coordinates (3, –4) and (–3, 10) both lie on the line with the equation $\boldsymbol{y = mx + b}$, can we find the equation for the line? Let's starting by finding the value of $\boldsymbol{b}$.

First, plug in the two coordinate pairs to make two equations:

$$y = mx + b$$

$$-4 = m(3) + b$$

$$10 = m(-3) + b$$

Then, evaluate as a system of equations. One method is to add the two equations together and see if one of the variables cancels out.

$$-4 + 10 = m(3) + b + m(-3) + b$$

$$6 = m(3) + m(-3) + b + b \quad \text{Combine like terms.}$$

$$6 = m(0) + 2b$$

$$6 = 2b \quad \text{m(0) cancels out.}$$

$$3 = b$$

Another method is to use the slope formula to solve for *m* first and then solve for *b*.

Slope Formula

To find the slope of a line containing the points (x_1, y_1) and (x_2, y_2):

$$\frac{y_1 - y_2}{x_1 - x_2}$$

REMEMBER...

Keep your coordinates in the same order for the top and bottom of the fraction, or else you'll get the wrong sign.

Plug in the points (3, –4) and (–3, 10) to solve for slope:

$$\frac{y_1 - y_2}{x_1 - x_2}$$

$$= \frac{10 - (-4)}{(-3) - 3}$$

$$= \frac{14}{-6}$$

$$= -\frac{14}{6}$$

$$= -\frac{7}{3}$$

Therefore, $m = -\frac{7}{3}$. Plug that back into the equation:

$$y = mx + b$$
$$y = -\frac{7}{3}x + b$$

Finally, plug in one of the given coordinate pairs for x and y, and solve for b:

$$y = -\frac{7}{3}x + b$$

$$-4 = -\frac{7}{3} \times 3 + b$$

$$-4 = -7 + b$$

$$-4 + 7 = b$$

$$3 = b$$

FIND THE INTERSECTION OF TWO LINES

If two lines intersect, they do so at a single point. To find that point, you can solve as a system of two equations.

Find the intersection of ($y = 4x - 3$) and ($y = 2x + 5$).

In the previous example, we solved a system of equations by adding the equations together. Another method for solving systems is called **substitution**—using an equivalent expression for a certain variable so that you can solve for another variable.

If each of the equations has y alone, that makes it easy, because we can just set the expressions as equal to each other. We know that $y = y$, so the two expressions are also equal to each other.

$$y = 4x - 3$$
$$y = 2x + 5$$
$$4x - 3 = 2x + 5$$

Now that the y variables are gone, we are able to solve for x.

$$4x - 3 = 2x + 5$$
$$4x = 2x + 5 + 3$$ Add 3 to both sides of the equation.
$$4x = 2x + 8$$
$$4x - 2x = 8$$ Subtract $2x$ from both sides of the equation.
$$2x = 8$$
$$x = 4$$ Divide both sides of the equation by 2.

We now have a value for x, but we still need y. Plug in $x = 4$ to either one of the original equations, and evaluate for y.

$$y = 4x - 3$$
$$y = 4 \times 4 - 3$$
$$y = 16 - 3$$
$$y = 13$$

Therefore, the intersection point is (4, 13).

Intersection points should always satisfy both equations, so you can check your work by plugging the coordinate pair into the other equation:

$$y = 2x + 5$$

$$13 = 2 \times 4 + 5$$

$$13 = 8 + 5$$

$$13 = 13 \quad \checkmark \text{ True!}$$

PARALLEL LINES

If a pair of linear equations has no solution, that means that the two lines do not intersect at any point. And we know that when two lines do not intersect, it means that the lines are **parallel.** This is an algebraic explanation for the slopes of parallel lines being equal. For example, $(2x + 4)$ can't possibly have a value equal to $(2x + 9)$; therefore, we know that the pair of equations $(y = 2x + 4)$ and $(y = 2x + 9)$ has no solutions.

Consider the pair of equations $\mathbf{y} + 9 = 7\mathbf{x}$ and $3\mathbf{y} = 21\mathbf{x} + 6$. If the pair of equations has a solution, we can solve algebraically using a system of equations.

Whenever you have a linear equation that's not already in the form of $y = mx + b$, a good first step is to put the equation in that form. This makes the equations much easier to compare.

For the first equation, y has a term added to it. To get y by itself, subtract 9 from both sides of the equation:

$$y + 9 = 7x$$

$$y = 7x - 9 \quad \text{Subtract 9 from both sides of the equation.}$$

For the second equation, y has a coefficient. To get y by itself, divide both sides of the equation by 3:

$$3y = 21x + 6$$

$$y = \frac{21x + 6}{3}$$ Divide both sides of the equation by 3.

$$y = 7x + 2$$ Simplify (distributive property).

Now, try setting the expressions equal to each other:

$$7x - 9 = 7x + 2$$

$$-9 = 2$$ Subtract $7x$ from both sides of the equation.

If you subtract $7x$ from both equations, you get $-9 = 2$, which is obviously not a true equation. In other words, there is no value for y that can equal *both* quantities $(7x - 9)$ and $(7x + 2)$. This means that there is no solution for this pair of equations—no (x, y) coordinate pair that satisfies both equations.

Slopes of Parallel Lines

If two lines are parallel, then they have the same slope, but different y-intercepts.

If two lines have the same slope, but different y-intercepts, then they are parallel.

Sometimes, you'll find that two equations are actually the same.

Suppose we need to determine the solution(s) for the lines $(2y = 6x - 7)$ and $(8y + 28 = 24x)$.

First, put both equations into the form of $y = mx + b$:

$$2y = 6x - 7$$

$$y = \frac{6x - 7}{2} \quad \text{Divide both sides of the equation by 2.}$$

$$y = 3x - \frac{7}{2} \quad \text{Simplify (distributive property).}$$

$$8y + 28 = 24x$$

$$8y = 24x - 28 \quad \text{Subtract 28 from both sides of the equation.}$$

$$y = \frac{24x - 28}{8} \quad \text{Divide both sides of the equation by 8.}$$

$$y = 3x - \frac{7}{2} \quad \text{Simplify (distributive property).}$$

Therefore, these two equations actually represent the same line. If you graph both equations, they overlap each other entirely. The two equations have an **infinite** number of solutions.

DISTANCE AND MIDPOINT

In this lesson, we'll review how to calculate the distance and midpoint from two points in the coordinate plane. You'll also learn how to divide a line to a certain ratio, such as 1:3.

DISTANCE

The **distance** between two coordinates is the measure of the length between them. In the coordinate plane, you'll determine distance by using calculations, rather than a measuring tool such as a ruler.

One straightforward way to calculate distance is to use the Pythagorean Theorem. In other words, you can construct a right triangle in which the two points form the hypotenuse. Then, use the Pythagorean Theorem to find the length of the hypotenuse.

There are two ways that you can draw such a right triangle from two points—think of it like two halves of a rectangle. It does not matter which way you draw the triangle, because both are rotations of each other and are therefore congruent. The hypotenuse will be the same either way.

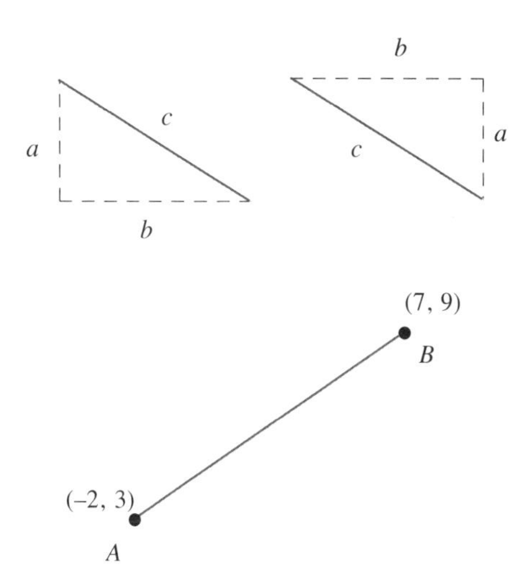

In this example, we will find the length of *AB*.

First, make a right triangle from the figure. Extend a horizontal line from one of the two points, and a vertical line from the other, to where they intersect.

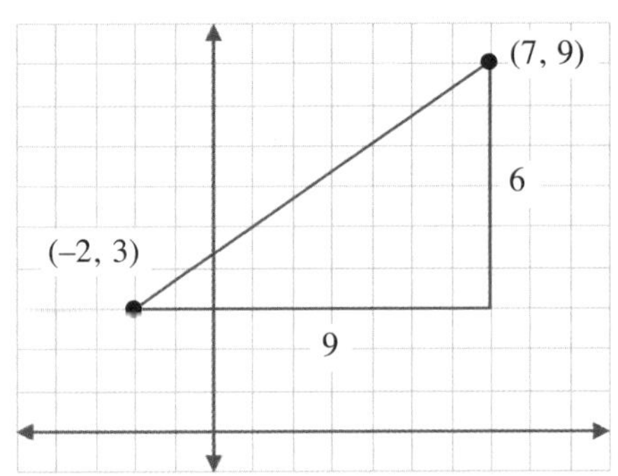

You can measure the legs of this triangle by counting, or by subtracting the *x* and *y* coordinates, respectively. The legs of this triangle are 6 and 9, as shown in the figure above.

Then, use the Pythagorean Theorem to solve for the hypotenuse:

$$a^2 + b^2 = c^2$$

$$6^2 + 9^2 = c^2$$

$$36 + 81 = c^2$$

$$117 = c^2$$

$$\sqrt{117} = c$$

The length of *AB* is $\sqrt{117}$, or ≈ 10.82.

DISTANCE FORMULA

For the distance formula, you do not have to worry about the order of $(x_1 - x_2)$ versus $(x_2 - x_1)$— since you end up squaring these values, the result will always be positive.

The distance formula is derived from the Pythagorean Theorem.

First, if we take the square root of both sides of the equation, we solve for *c*:

$$a^2 + b^2 = c^2$$

$$\sqrt{a^2 + b^2} = c$$

We can also substitute for *a* and *b* using the coordinates of the two points:

$$a = x_1 - x_2 \qquad\qquad b = y_1 - y_2$$

Substitute those values for *a* and *b*, respectively:

$$c = \sqrt{a^2 + b^2}$$

$$c = \sqrt{(x_1 - x_2)^2 + (y_1 - y_2)^2}$$

That's the distance formula—or, you could call it "Pythagorean Theorem in the Coordinate Plane."

Distance Formula

To find the distance *d* between two points (x_1, y_1) and (x_2, y_2):

$$d = \sqrt{(x_1 - x_2)^2 + (y_1 - y_2)^2}$$

Try plugging the points from the previous example, (–2, –3) and (7, 9), into the distance formula:

$$d = \sqrt{(x_1 - x_2)^2 + (y_1 - y_2)^2}$$

$$d = \sqrt{(-2 - 7)^2 + (3 - 9)^2}$$

$$d = \sqrt{(-9)^2 + (-6)^2}$$

$$d = \sqrt{81 + 36}$$

$$d = \sqrt{117}$$

$$d \approx 10.82$$

This is the same result that we got when we used the Pythagorean Theorem.

Memorizing the distance formula may help make your calculations easier, but either method is valid.

MIDPOINT

The **midpoint** of two coordinates is the point that lies exactly halfway between them.

Consider what you do when you need to find the midpoint of two *numbers*—you just take the average of the two numbers. For example, we know that 6 is exactly halfway between 2 and 10, because 6 is the average of 2 and 10: $\frac{2+10}{2} = 6$.

To find the midpoint between two coordinates, you're essentially doing the same thing—finding the average of the numbers. The difference is that there are two pairs of numbers instead of one. More specifically, you're finding the average of the *x*-coordinates and then the average of the *y*-coordinates.

Another way to think about midpoint is "the point that is halfway between the two *x*-values, and halfway between the two *y*-values."

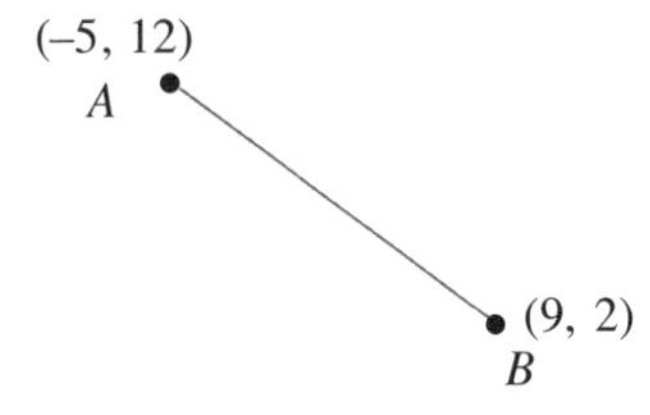

To calculate the midpoint between these coordinates, find the average of the *x*-values; then find the average of the *y*-values.

The average of the x-coordinates is $\frac{-5+9}{2}$, which equals 2. This is the x-coordinate of the midpoint.

The average of the y-coordinates is $\frac{12+2}{2}$, which equals 7. This is the y-coordinate of the midpoint.

Therefore, the midpoint is (2, 7).

MIDPOINT FORMULA

The Midpoint Formula is derived from taking the average of the x- and y-coordinates, respectively.

Midpoint Formula

To find the midpoint m between two points (x_1, y_1) and (x_2, y_2):

$$\text{midpoint} = \left(\frac{x_1 + x_2}{2}, \frac{y_1 + y_2}{2}\right)$$

PARABOLAS

A **parabola** is a U-shaped curve. It belongs to a group of objects called **conic sections**—literally, cross-sections of cones. A parabola is also what you get when you graph a **quadratic equation**—an equation in which one of the terms is raised to a power of 2 (for example, $y = x^2 + 2x + 4$).

The **vertex** is the point at which the parabola changes direction—that is, the curve swings from down to up, or left to right. All parabolas have a line of symmetry—called the **axis of symmetry**—that passes through the vertex.

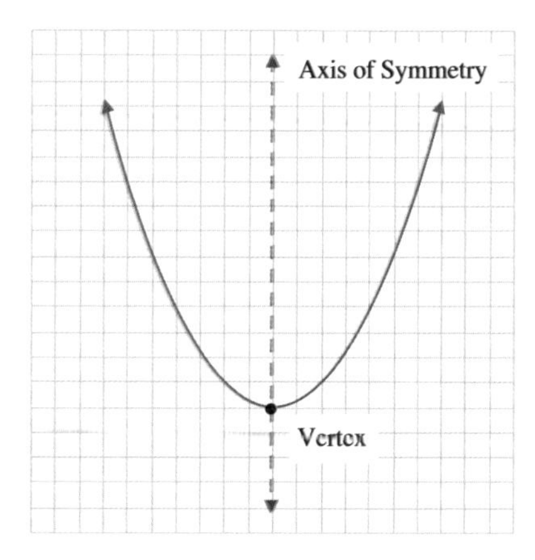

The definition of a parabola involves a point, called the **focus**, and a line, called the **directrix.** The parabola is the set of points that are equidistant from the focus and the directrix. In other words, if you have a point on a parabola (call the point *X*), the straight-line distance between *X* and the focus is equal to the perpendicular distance between *X* and the directrix.

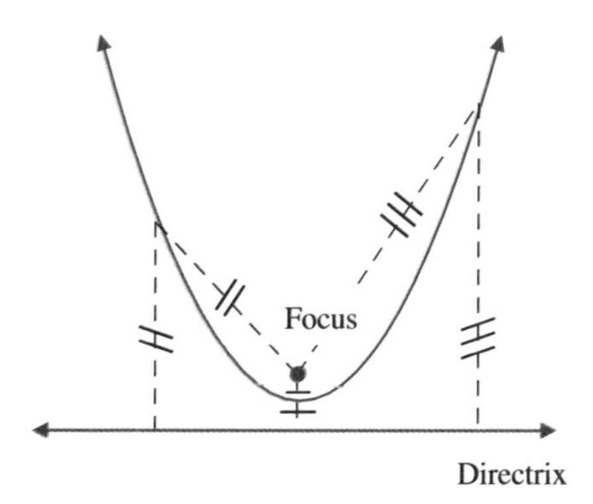

Note that the axis of symmetry of a parabola is perpendicular to the directrix and passes through the focus and vertex. The vertex is always halfway between the focus and directrix.

GRAPHING A PARABOLA FROM AN EQUATION

Equation of a Parabola: Standard Form

$y = ax^2 + bx + c$ (vertical parabola)

$x = ay^2 + by + c$ (horizontal parabola)

You may have had some experience graphing parabolas in algebra class. One standard method is to first factor the quadratic equation, and then solve for the **roots** (also known as **zeroes**) of the equation—that is, the x-values that make $y = 0$. This gives two points on the parabola, which you can then use to find the vertex. For the purposes of this lesson, we'll assume that you're somewhat familiar with factoring quadratic expressions. If not, you may find it helpful to brush up a little before proceeding.

For this example, let us graph the parabola represented by the equation $\mathbf{y = x^2 + 8x + 15}$.

Try factoring the quadratic expression. Begin by finding the factors of 15:

1, 15

3, 5

Now, find the pair of factors that adds up to 8.

$1 + 15 = 8?$ No.

$3 + 5 = 8?$ Yes!

That means that the binomial factors of $x^2 + 8x + 15$ are $(x + 3)$ and $(x + 5)$.

$$y = x^2 + 8x + 15$$

$$y = (x + 3)(x + 5)$$

Now, find the values for x where $y = 0$.

$$0 = (x + 3)(x + 5)$$

That means that either $(x + 3)$ or $(x + 5)$ equals zero.

$$\text{If } (x + 3) = 0, \text{ then } x = -3.$$

$$\text{If } (x + 5) = 0, \text{ then } x = -5.$$

Plugging in either of those two values into the original equation will make $y = 0$.

$$\begin{aligned} y &= (x + 3)(x + 5) \\ &= (-3 + 3)(-3 + 5) \\ &= (0)(2) \\ &= 0 \end{aligned}$$

and

$$\begin{aligned} y &= (x + 3)(x + 5) \\ &= (-5 + 3)(-5 + 5) \\ &= (-2)(0) \\ &= 0 \end{aligned}$$

Therefore, the two roots of the equation are $x = -3$ and $x = -5$. In other words, the coordinates $(-3, 0)$ and $(-5, 0)$ exist on the parabola, where the parabola intersects the x-axis.

The parabola's vertex will have an x-value that is exactly halfway between the two zeroes. (This is also true using any two points that have the same y-value on the parabola.) The coordinate that's exactly between $(-3, 0)$ and $(-5, 0)$ is $(-4, 0)$.

Therefore, the x-coordinate of the vertex is -4. To find the corresponding y-value, plug -4 into the equation.

$$\begin{aligned} y &= (x + 3)(x + 5) \\ y &= (-4 + 3)(-4 + 5) \\ y &= (-1)(1) \\ y &= -1 \end{aligned}$$

Therefore, when $x = -4$, $y = -1$. This gives us the coordinate (–4, –1), which is the parabola's vertex. With the vertex and two zeroes, we can graph the parabola:

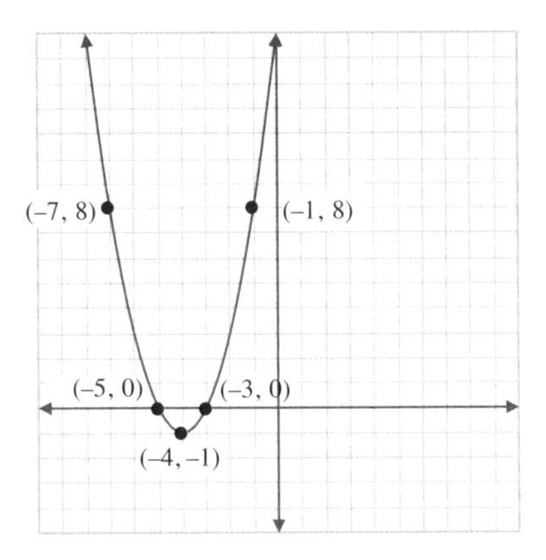

You may feel more comfortable with your graph if you plug in to find a few more points. For example, if $x = -1$, then $y = 8$ $((-1 + 3)(-1 + 5))$; and if $x = -7$, then $y = 8$ $((-7 + 3)(-7 + 5))$.

Equation of a Parabola: Vertex Form

$y = a(x - h)^2 + k$ (vertical parabola)

$x = a(y - k)^2 + h$ (horizontal parabola)

An equation in **vertex form** provides specific information—namely, the coordinate of the vertex. If you see an equation in this form, you can know immediately that the vertex is (h, k).

However, the vertex by itself is not enough to tell us the shape of the parabola. To plot the parabola, you'll need at least three points. Therefore, choose a couple of different values for x and find the corresponding y-coordinates.

In this example, we will graph a parabola whose equation is in vertex form: $y = 2(x - 4)^2 + 3$.

Since this equation is in vertex form, you know that the vertex of the parabola is (4, 3).

To graph the rest of the parabola, you'll need a couple more points. To find additional points that lie on the parabola, you can choose different values for x and then evaluate for the corresponding y-coordinates.

Let's try $x = 2$.

$$y = 2(x - 4)^2 + 3$$

$$y = 2(2 - 4)^2 + 3 \quad \text{Plug in } x = 2.$$

$$y = 2(-2)^2 + 3 \quad \text{Simplify.}$$

$$y = 2(4) + 3$$

$$y = 8 + 3$$

$$y = 11$$

Therefore, the point (2, 11) lies on the parabola.

Try one additional value for x. We'll use $x = 6$.

$$y = 2(x - 4)^2 + 3$$

$$y = 2(6 - 4)^2 + 3 \quad \text{Plug in } x = 6.$$

$$y = 2(2)^2 + 3 \quad \text{Simplify.}$$

$$y = 2(4) + 3$$

$$y = 8 + 3$$

$$y = 11$$

Therefore, the point (6, 11) also lies on the parabola.

With three points, the parabola can be graphed:

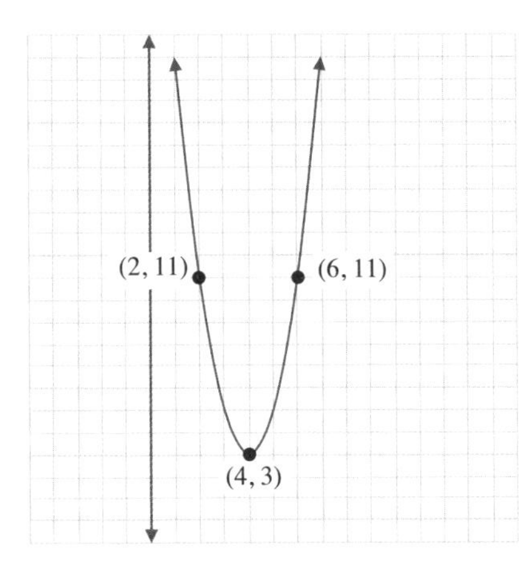

You can also derive the focus and directrix from the vertex form of an equation.

Focus and Directrix from Vertex Form

For a parabola with vertex (h, k):

$y = \frac{1}{4p}(x - h)^2 + k$ (vertical parabola)

$x = \frac{1}{4p}(y - k)^2 + h$ (horizontal parabola)

The value $\frac{1}{4p}$ is equal to the distance between the vertex and the focus. This is also equal to the distance between the vertex and the directrix.

Therefore, if you have a parabola in vertex form, you can add the quantity $\frac{1}{4a}$ to the vertex coordinate in order to find the focus.

Likewise, you can subtract $\frac{1}{4a}$ from the vertex coordinate to find the y-coordinate of the directrix (for a vertical parabola).

For instance, we can find the vertex and directrix for the parabola with the equation $y = \left(\frac{1}{8}\right)(x - 4)^2 + 3$.

Since this equation is already in vertex form, you know that $a = \frac{1}{8}$. Find the quantity $\frac{1}{4a}$:

$$\frac{1}{4a} = \frac{1}{4\left(\frac{1}{8}\right)} = \frac{1}{\frac{4}{8}} = \frac{1}{\frac{1}{2}} = 2$$

Therefore, the distance between the vertex and the focus is 2. The vertex is (4, 3), from the equation. Since this is a vertical parabola (y is isolated in the equation), just add 2 to the y-coordinate.

Vertex = (4, 3)

Focus = (4, 5)

The directrix, then, will be two units below the vertex, so it passes through the coordinate (4, 1). Remember that the directrix is a line, which in this case is the line $y = 1$.

Directrix: $y = 1$

FINDING THE EQUATION OF A PARABOLA

FIND THE EQUATION, GIVEN THE ZEROES

To find the equation of a parabola from its graph, try to start with the zeroes—that is, the points where the parabola crosses the x-axis. This will give you two factors of the equation. For the purposes of this exercise, let's say that we have two points on the parabola at (m, 0) and (n, 0). Then, we know that two factors are $(x - m)$ and $(x - n)$.

However, that's not enough to derive the full equation, because there are actually an infinite number of parabolas that can pass through those two points.

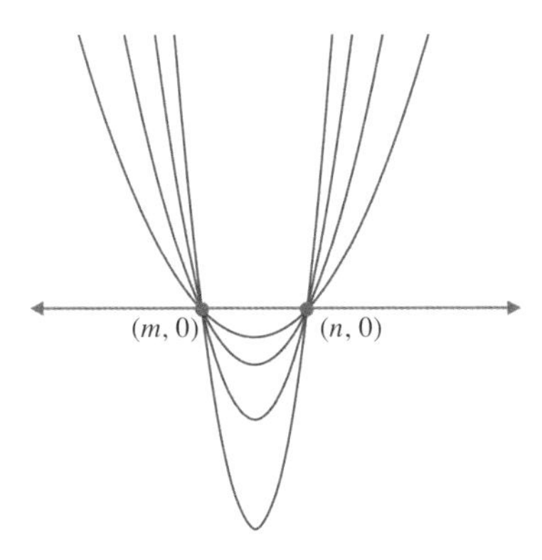

Therefore, we need a third point in order to find the missing factor, a. You're going to put your equation in the following form, and plug in the values of a known coordinate to solve for a.

$$y = a\,(x - m)(x - n)$$

In this example, we will find the equation of the parabola with zeroes at (–5, 0) and (3, 0), and the vertex (–1, –4).

If the zeroes are (–5, 0) and (3, 0), then you know that two factors of the equation are $(x + 5)$ and $(x - 3)$.

$$y = a\,(x + 5)(x - 3)$$

To solve for a, plug in the coordinates of the vertex (that is, $x = -1$ and $y = -4$).

$$\begin{aligned}
y &= a\,(x + 5)(x - 3) & \\
-4 &= a\,(-1 + 5)(-1 - 3) & \text{Plug in } x = -1 \text{ and } y = -4. \\
-4 &= a\,(4)(-4) & \text{Simplify.} \\
-4 &= a\,(16) & \\
\frac{-4}{16} &= a & \text{Divide both sides of the equation by 16.} \\
-\frac{1}{4} &= a & \text{Simplify.}
\end{aligned}$$

Therefore, the equation of the parabola is $y = \frac{1}{4}(x + 5)(x - 3)$.

FIND THE EQUATION, GIVEN THE FOCUS AND DIRECTRIX

Recall that any point on a parabola is equidistant from the focus and the directrix. That is, if you have a point (x, y), then the distance to the focus is equal to the distance to the directrix. Therefore, to derive the equation of a parabola, we can just find those two distances and set them equal to each other.

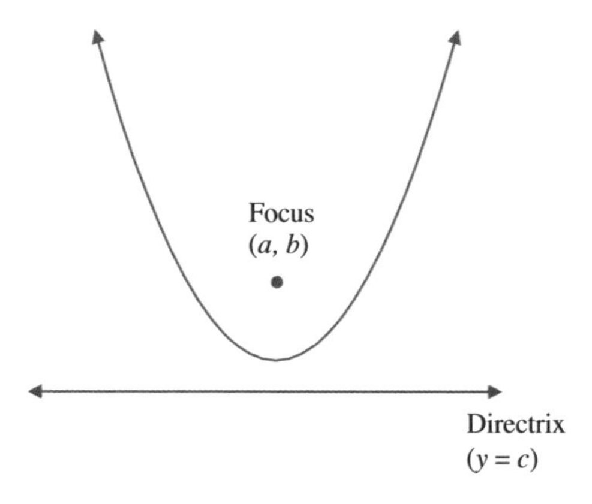

In the figure above, the focus of the parabola is (a, b) and the directrix is the line $y = c$. Take a point on the parabola, (x, y). To calculate the distance from (x, y) to (a, b), use the distance formula or Pythagorean Theorem:

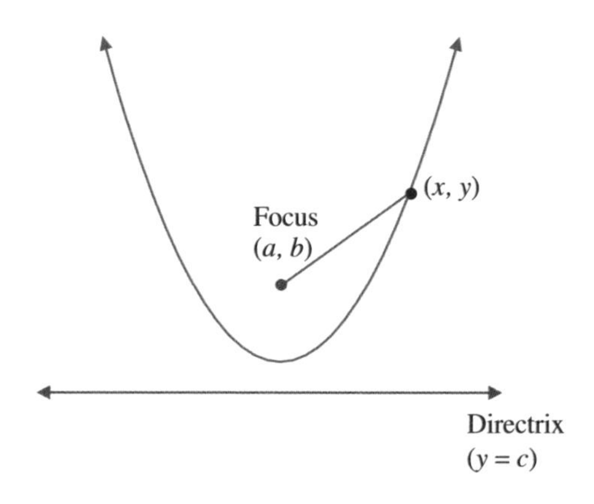

Distance from (x, y) to (a, b):

$$d = \sqrt{(x-a)^2 + (y-b)^2}$$

Now, find the distance between (x, y) and the directrix. Recall that the shortest distance between a point and a line will be through a segment perpendicular to the line, as shown:

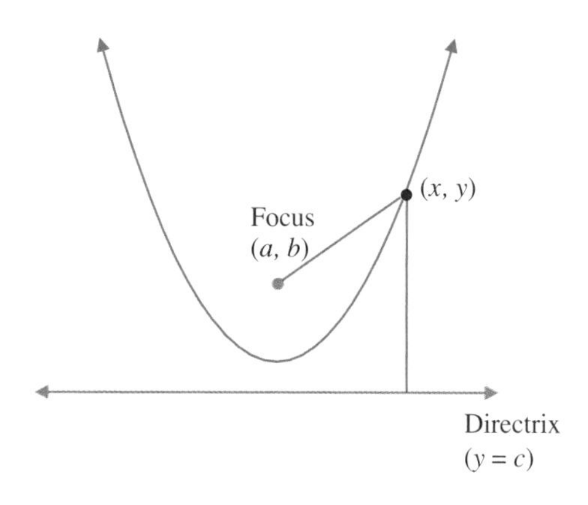

Therefore, the distance between (x, y) and the directrix is $|y - c|$. (Use absolute value, because distances are always positive.)

Now, set the values equal to each other, and simplify. Yes, this will have a lot of steps! Work carefully, and don't do anything in your head.

$$|y-c| = \sqrt{(x-a)^2 + (y-b)^2}$$

$(y-c)^2 = \left(\sqrt{(x-a)^2 + (y-b)^2}\right)^2$	Square both sides of the equation.
$(y-c)^2 = (x-a)^2 + (y-b)^2$	The root sign cancels out.
$y^2 - 2yc + c^2 = (x-a)^2 + (y-b)^2$	Expand $(y-c)^2$.
$y^2 - 2yc + c^2 = (x-a)^2 + y^2 - 2yb + b^2$	Expand $(y-b)^2$.
$y^2 + 2yb - 2yc + c^2 = (x-a)^2 + y^2 + b^2$	Add $2yb$ to both sides.
$y^2 + 2yb - 2yc = (x-a)^2 + y^2 + b^2 - c^2$	Subtract c^2 from both sides.
$2yb - 2yc = (x-a)^2 + b^2 - c^2$	Subtract y^2 from both sides.
$2y(b-c) = (x-a)^2 + b^2 - c^2$	Factor $2y$ from the left side of the equation.
$y = \frac{(x-a)^2 + b^2 - c^2}{2(b-c)}$	Divide both sides of the equation by $2(b-c)$.

That's quite a bit of algebra, there, and it can still be simplified further! But this format is fine, and it has y by itself.

Equation of a Parabola, with Focus and Directrix

Given focus (a, b) and directrix $y = c$:

$$y = \frac{(x-a)^2 + b^2 - c^2}{2(b-c)}$$

CHAPTER 8
TRANSFORMATIONS

© istockphoto.com / 18percentgrey

In everyday language, transformation means change. In geometry, transformation refers to changing a figure. You can transform a figure by moving, flipping, turning, or resizing it.

CHAPTER CONTENTS

INTRO TO TRANSFORMATIONS

Examples of Transformations

Two figures are **congruent** if they have the same shape and size. If a figure is moved, flipped, or turned, the original and resulting figures are still congruent. These types of transformations are called **rigid motions**—the location or position of the figure may change, but its shape and size are the same. In other words, lengths and angles of the figure remain the same. Another word for rigid motion is **isometry**.

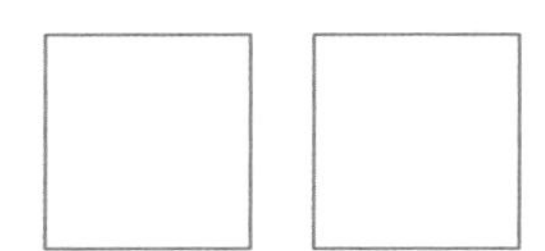

Example of Congruent Figures

When discussing transformations, we refer to the original figure as the **pre-image**, and the new, transformed figure as the **image**. In the figure above, the image and pre-image are congruent.

Two figures are **similar** if they have the same shape, but they may have different sizes. If a figure is proportionally resized, the original and resulting figures are similar, but not congruent.

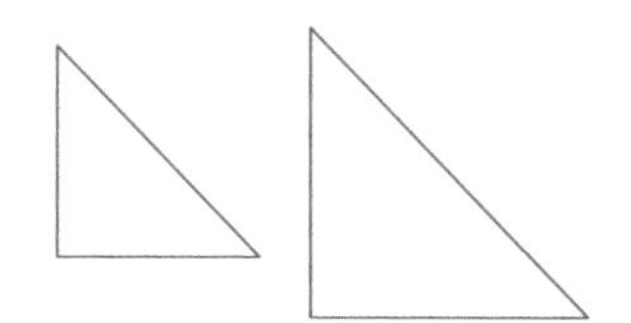

Example of Similar Figures

TRANSLATION

Translation refers to moving a figure to a new location. In other words, it's shifting the figure up, down, left, right, or diagonally.

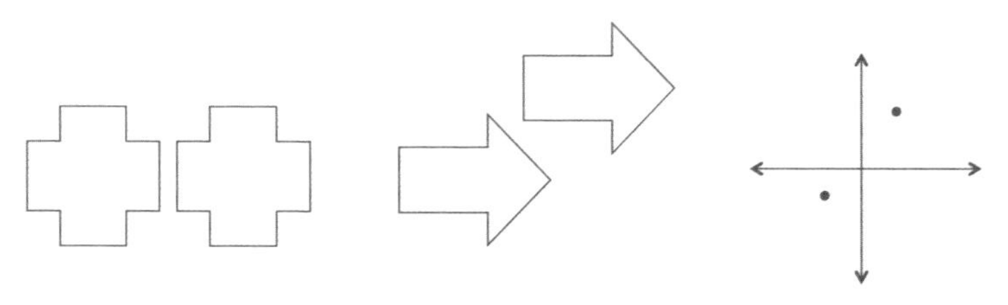

Examples of Translations

The most common way of dealing with translations is in the coordinate plane. You may need to move a figure in a plane, identify how a figure was moved, or write an algebraic expression for a translation.

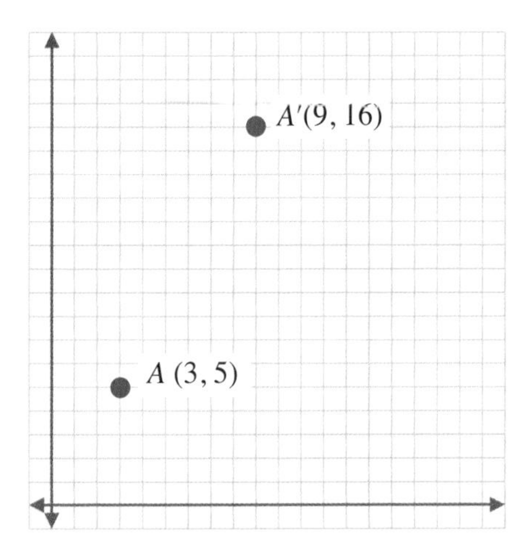

When labeling points, it's customary to use the same letter for the image and pre-image, and usually to use capital letters. You should also add a **prime mark** to the label for the image, to distinguish it from the pre-image. A prime mark is almost like an apostrophe, but straighter, and it's used almost exclusively for mathematics. In the figure above, A is the pre-image and A' is the image.

> Use a prime mark to distinguish images from pre-images. Example: Pre-image A and image A'

> The prime mark is also used to label feet and inches. A single prime mark is used for feet, and a double prime mark for inches. For example, 5'7" means 5 feet and 7 inches.

On the graph, A' is the result of moving A upward and to the right. To find out how much it moved, find the difference between the two coordinates. When looking at the x-coordinates, we see that it moved from 3 to 9. The difference between these coordinates is 6 ($9 - 3 = 6$); so, this translation moved the point to the right by 6 coordinates. Now do the same for the y-coordinates. The difference between the y-coordinates is 11 ($16 - 5 = 11$); so, this translation moved the point upward by 11 coordinates.

To describe this translation, you can say that the point has been moved horizontally 6 units, and vertically 11 units. (It makes more sense to describe the *x* first, since *x* comes first in *x*, *y* coordinate pairs.)

You can also describe the translation in algebraic terms. The algebraic notation for a translation uses the letter *T*. For example, $T(\Delta ABC) = \Delta A'B'C'$ means that a triangle *ABC* has been translated to new coordinates *A'B'C'*. You can also indicate the distance of the translation. For example, $T_{(2, 5)}(\Delta ABC)$ tells us that the triangle is translated 2 units right, and 5 units up.

Notating Translations

$(x + a)$ moves a point *a* units in the positive *x* direction (right).

$(x - a)$ moves a point *a* units in the negative *x* direction (left).

$(y + a)$ moves a point *a* units in the positive *y* direction (up).

$(y - a)$ moves a point *a* units in the negative *y* direction (down).

Now take a look at this graph.

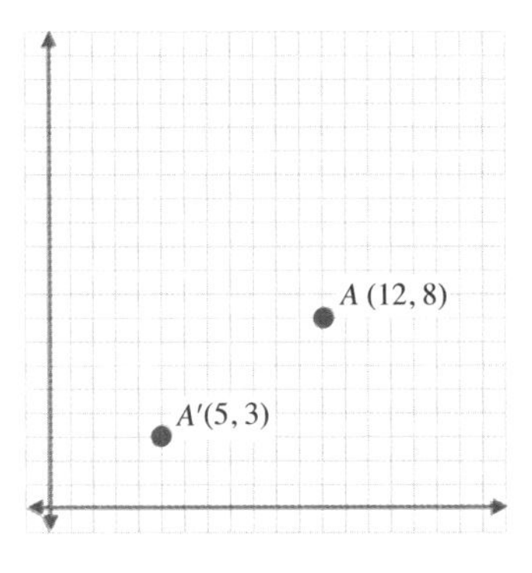

In this example, the point has moved down and to the left. The translation is negative in both directions. We can find the difference between the coordinates, just as we did in the previous example. The difference between the x-coordinates is $(5 - 12)$, or -7. The difference between the y-coordinates is $(3 - 8)$, or -5.

Therefore, this translation is horizontally -7, vertically -5.

We can also express this translation algebraically: $(x - 7, y - 5)$

REMEMBER...

If you're not sure about signs in the coordinate plane,
down is negative, and up is positive;
left is negative, and right is positive.

TRANSLATING FUNCTIONS

In Algebra, you may have learned a little bit about graphing functions. We'll use basic examples in this section, so for now, don't worry about remembering everything you learned in class. However, you should know that a function takes an input and produces an output. For example, in the function $f(x) = x + 5$, x is the input, and $x + 5$ is the output. So, if $x = 1$, the output is $1 + 5 = 6$, and if $x = 3$, the output is $3 + 5 = 8$.

REMEMBER...

When working with graphs of functions,
input values are graphed on the x-axis,
and output values are graphed on the y-axis.

In the examples that follow, you'll see that by changing a function, you can observe fairly predictable changes in the graph of that function.

Consider the function $f(x) = x^2$. In this function, we input a value x and get a value of x^2 as the output. Some examples of points for this function would be $f(2) = 4$, $f(3) = 9$, and so on.

If we graph this function, the graph forms a curve. For every point on this curve, the value of the y-coordinate is the square of the value of the x-coordinate.

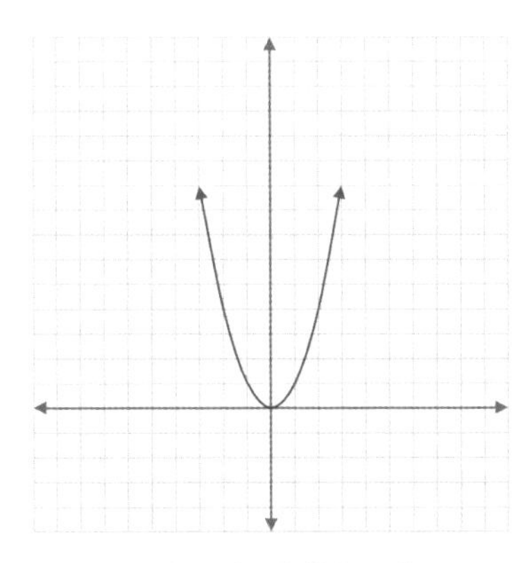

Graph of $f(x) = x^2$

If we change the function slightly, we'll change the appearance of the graph. Let's make a new function to relate to $f(x)$, which we previously defined as $f(x) = x^2$. We'll call the new function $g(x)$, and define it as $g(x) = f(x) + 4$.

What that means is that for every input x, we find the value of $f(x)$ and add 4 to it.

Let's plug in $x = 0$.

$$f(0) = 0$$
$$g(0) = f(0) + 4$$
$$g(0) = 0 + 4$$
$$g(0) = 4$$

This creates the point (0, 4). Some other examples of points in this function are (1, 5), (2, 8), and (–1, 5).

The effect on the graph is that it looks the same as $f(x)$, only it's 4 units higher. It's been translated upward.

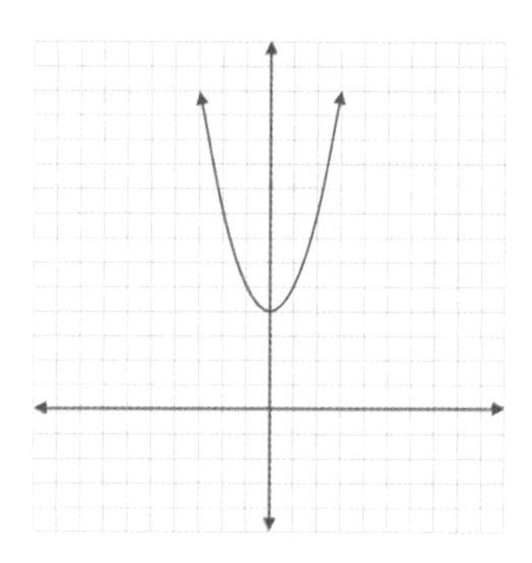

Graph of $g(x) = f(x) + 4$

Similarly, if we graph a function that equals $f(x) - 4$, we take the graph of $f(x) = x$ and shift it 4 units downward.

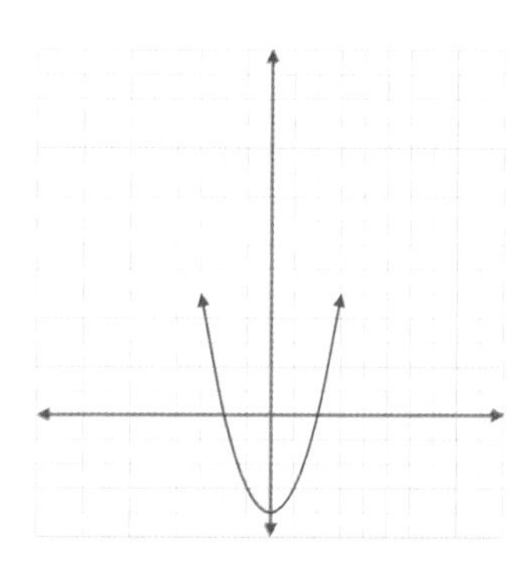

Graph of $g(x) = f(x) - 4$

If you change the expression used for the *input*, the graph will move differently. For example, let's say that $f(x) = x^2$, and a new function $h(x)$ equals $f(x + 4)$. If you plug in a value for x, what you're doing is adding $(x + 4)$ first and *then* inputting that value to $f(x)$.

Let's plug in $x = -4$.

$$h(-4) = f(-4 + 4)$$
$$h(-4) = f(0)$$
$$f(0) = 0$$
$$h(-4) = 0$$

Some other points on this graph are (–3, 1), (–5, 1), (–2, 4), (–6, 4), and so on.

The effect on the graph is that it looks the same as $f(x)$, only it's 4 units to the left. The $h(x)$ function is outputting the $f(x)$ values from 4 units over.

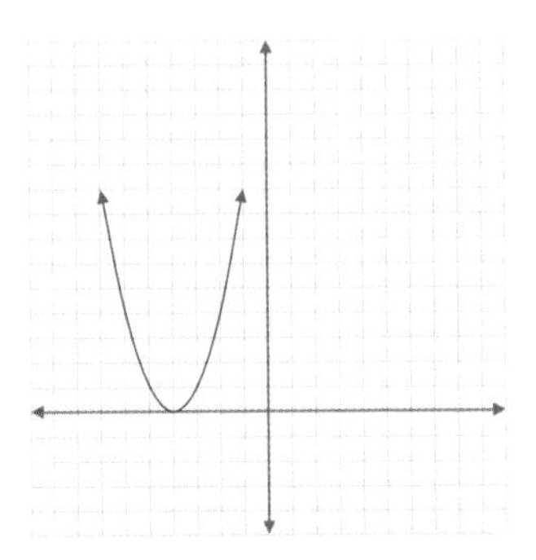

Graph of $h(x) = f(x + 4)$

Similarly, if we graph a function that equals $f(x - 4)$, we take the graph of $f(x) = x$ and shift it 4 units to the right.

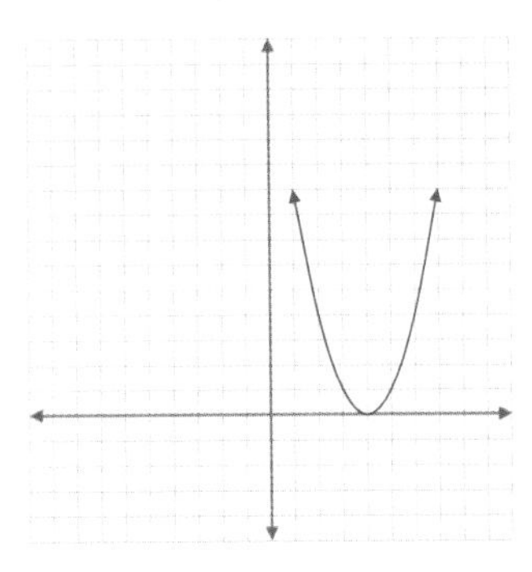

Graph of $h(x) = f(x - 4)$

Thus, the behavior of graphs follows a set of predictable patterns. To make working with graphs easier, try memorizing the following rules for translations:

Translation Rules for Functions

Compared to the graph of $y = f(x)$,

$y = f(x) + C$ moves the graph C units up

$y = f(x) - C$ moves the graph C units down

$y = f(x + C)$ moves the graph C units left

$y = f(x - C)$ moves the graph C units right

REFLECTION

Reflection is the term for a flipped version of an image. The way we think about reflection in real life, with mirrors, is very much the same way that reflection works in geometry. The reflected image is the same shape and size as the pre-image, but it's backwards.

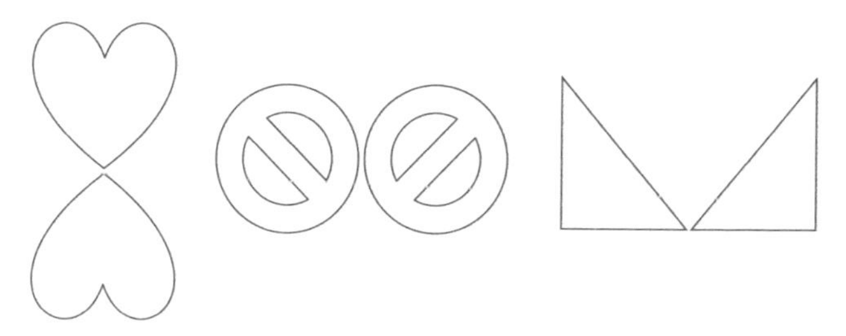

Examples of Reflections

As with translations, reflections are usually worked in the coordinate plane. Common exercises include drawing a reflection, identifying a reflection line, or writing an algebraic expression for a reflection.

One way to create a reflection is to use a folded piece of paper. Draw a figure on one side of the paper; then flip the folded paper over, and trace the image on the other side of the fold. The two figures will be congruent, but flipped. The fold in the paper serves as the **line of reflection**—the line across which the image is flipped.

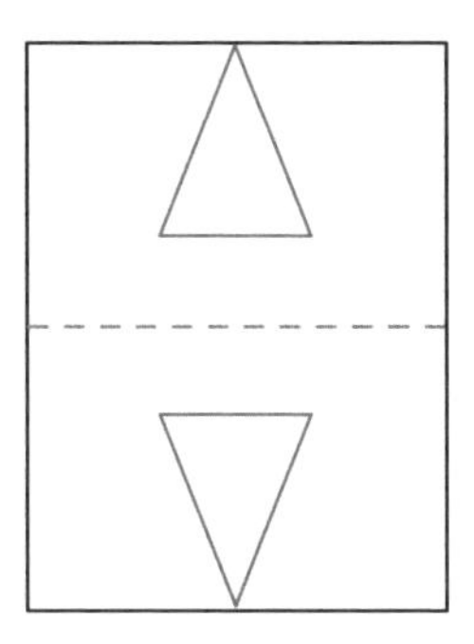

If you draw straight line segments between each pair of reflected points, as shown above, those segments will be parallel to each other. Additionally, the line of reflection forms a **perpendicular bisector** (a line that is perpendicular to another line segment, and intersects at its midpoint) with each of these segments. Each point in the pre-image is the same distance from the reflection line as its mirror image point.

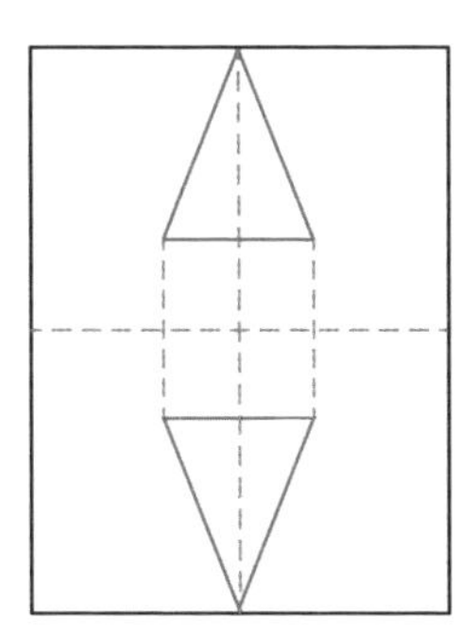

In the image below, point *A'* is the result of reflecting point *A* across the *y*-axis. That is, the reflection line is the *y*-axis itself.

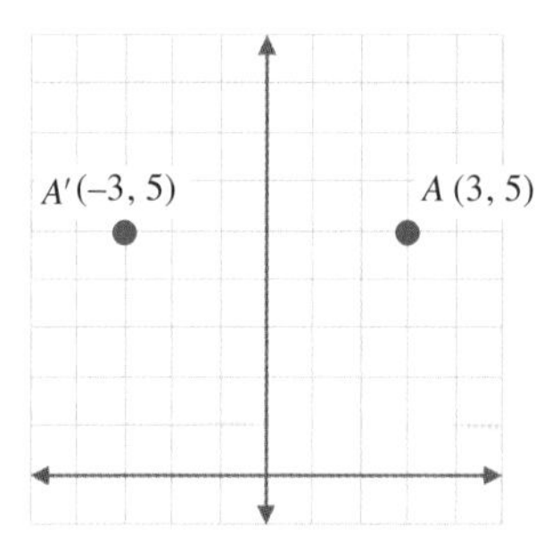

When reflecting a point across an axis, move straight across—that is, perpendicular to—the axis. Make sure the image and pre-image point are the same distance from the axis.

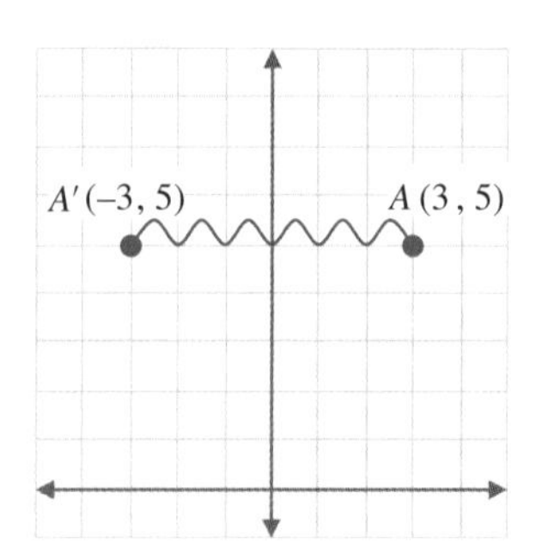

Reflection Across the Axis

When reflecting across the *y*-axis, the effect is that the *x*-coordinates will be the opposite of their pre-image.

$$R_{y=0}(x, y) = (-x, y)$$

When reflecting across the *x*-axis, the effect is that the *y*-coordinates will be the opposite of their pre-image.

$$R_{x=0}(x, y) = (x, -y)$$

REFLECTING FUNCTIONS

To recap, if you're reflecting across the x-axis, you're changing the sign of each of the y-coordinates of the figure. If you're reflecting across the y-axis, you're changing the sign of each of the x-coordinates of your figure.

The following graph shows the function $f(x) = x + 3$. We will reflect the function across the y-axis.

To reflect a function that forms a line, you can choose a few coordinate pairs from the line, and apply the reflection rules to those points. For example, what is $f(2)$ on this function? $f(2) = 2 + 3$, so $f(2) = 5$. That gives us the coordinate pair (2, 5). Great!

To reflect the coordinate (2, 5) across the y-axis, count over, or just change the sign of the x-coordinate. The reflected point is (–2, 5).

We'll need at least one more coordinate to make a line. How about $f(-2)$? $f(-2) = -2 + 3 = 1$. That gives us the coordinate pair (–2, 1). To reflect that coordinate across the y-axis, count over, or just change the sign of the x-coordinate. The reflected point is (2, 1).

Try a few more points for practice! Then, draw a line between your reflected points.

The reflected line looks like this:

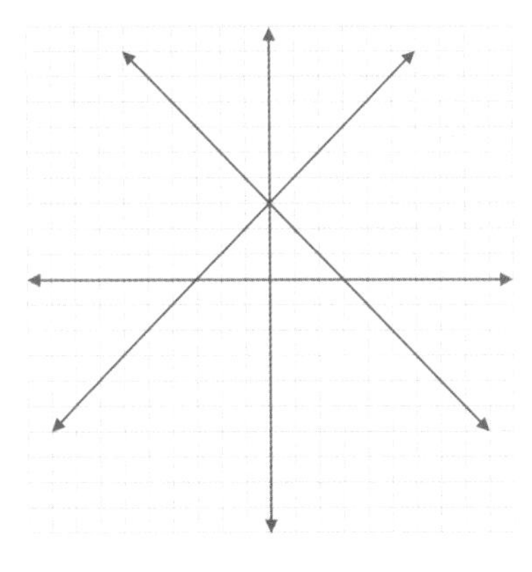

The rules for reflecting across the x-axis and y-axis can be expressed as follows:

Reflection Rules for Functions

Compared to the graph of $y = f(x)$,

$y = -f(x)$ reflects $f(x)$ across x-axis

$y = f(-x)$ reflects $f(x)$ across y-axis

ROTATION

In everyday language, "rotation" means turning or spinning around, such as with a wheel. In geometry, **rotation** means turning a figure around a fixed point. That point is called the **center of rotation.**

You may need to draw a rotation of a figure, identify the angle of rotation, or write an algebraic expression for a rotation.

Examples of Rotations

You can imagine a rotation as a pair of hands on a clock. The center of the clock, where both hands are attached, is like the **center of rotation**. The numbers that the hands point to are analogous to the image and pre-image points of a figure.

The figure above illustrates that there is a 60° angle between the 12 and 2. In fact, you can measure the angle from two different directions: clockwise or counterclockwise. (In the example above, the angle measured counterclockwise would be 60°, and the same angle measured clockwise would be –300°.) Both measurements are valid; however, it's important to remember the difference in signs.

Rotations are **positive** when measured in the counterclockwise direction.

Rotations are **negative** when measured in the clockwise direction.

When you're given a corresponding image and pre-image point, you can measure the **angle of rotation** by drawing the angle between them. More specifically, draw a line segment from the image point to the center of rotation, then another segment from the pre-image point to the center of rotation. The two line segments form an angle, which you can then measure with a protractor.

"Clockwise" is the direction of the hands of a clock (i.e., as the hands move from 1 to 2 to 3, and so on). "Counterclockwise" is the opposite direction.

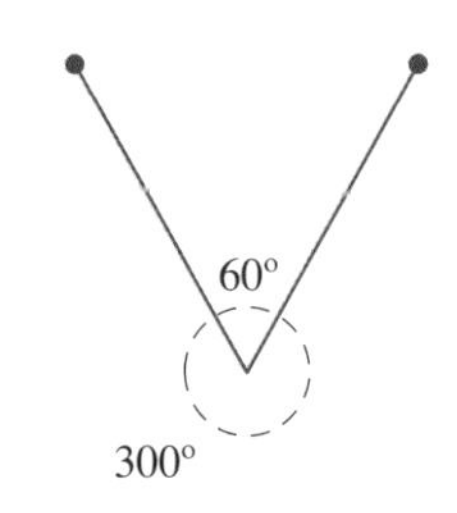

You can use the same approach for rotations of 180°, 270°, or 360°, since these angles are multiples of 90°. 180° is two turns of 90°:

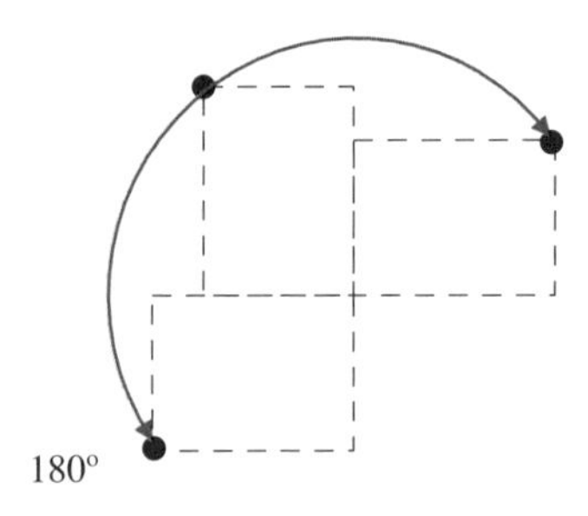

270° is three turns of 90°:

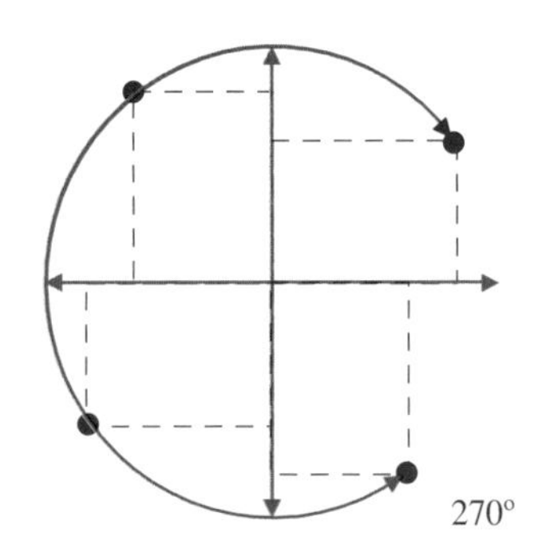

360° is four turns of 90°:

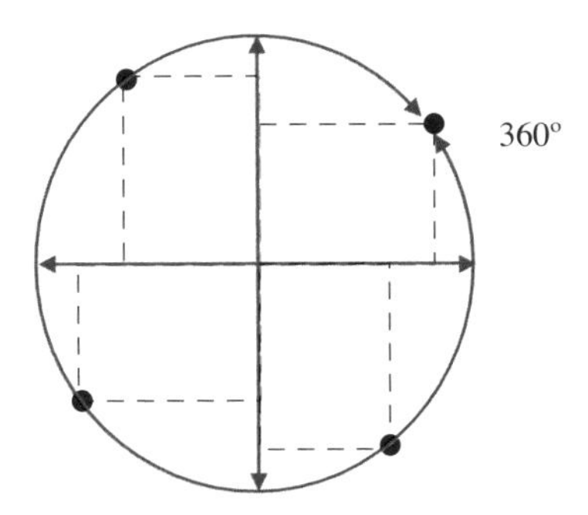

For rotations of 90° and multiples of 90°, use the following rules (note that these rules work only when the center of rotation is the origin):

Rotation Rules for Points

If the center of rotation is the origin O (0, 0), then:

$$r_{(90°, O)}(x, y) = (-y, x)$$
$$r_{(180°, O)}(x, y) = (-x, -y)$$
$$r_{(270°, O)}(x, y) = (y, -x)$$
$$r_{(360°, O)}(x, y) = (x, y)$$

SYMMETRY

A figure has **symmetry** if, after certain transformations, the image is identical to the pre-image, and in the same position. There are different kinds of symmetry, including reflectional symmetry and rotational symmetry. Some figures have no symmetry, some have exactly one kind of symmetry, while others may have multiple kinds of symmetry.

REFLECTIONAL SYMMETRY

A figure has **reflectional symmetry** when one half of the image is the mirror image of the other half. In other words, the reflected image just overlaps itself. This is also known as **line symmetry** or **mirror symmetry**. When people talk about "symmetry" in everyday language, they're most likely referring to reflectional symmetry.

Examples of Figures with Reflectional Symmetry

A **line of symmetry** is the line of reflection that results in symmetry.

Think of a line of symmetry as a fold line in a piece of paper. If you can fold a shape and have two identical halves that line up with each other, the shape has line symmetry. A figure can have no lines of symmetry, one line of symmetry, or multiple lines of symmetry.

A rectangle has two lines of symmetry. If you fold it in half vertically or horizontally, you produce two identical halves that line up with each other.

However, if you fold the rectangle in half diagonally, the two halves do not match up, even though they are congruent. This is NOT a line of symmetry.

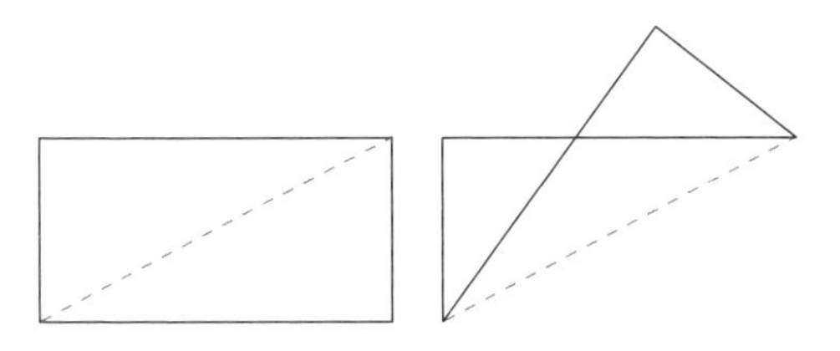

A square has 4 lines of symmetry.

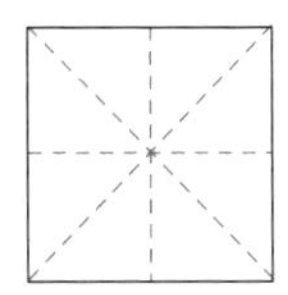

Additionally, any **regular polygon** has the same number of symmetry lines as it does sides. A regular pentagon has 5 lines of symmetry, a regular hexagon has 6 lines of symmetry, and so on.

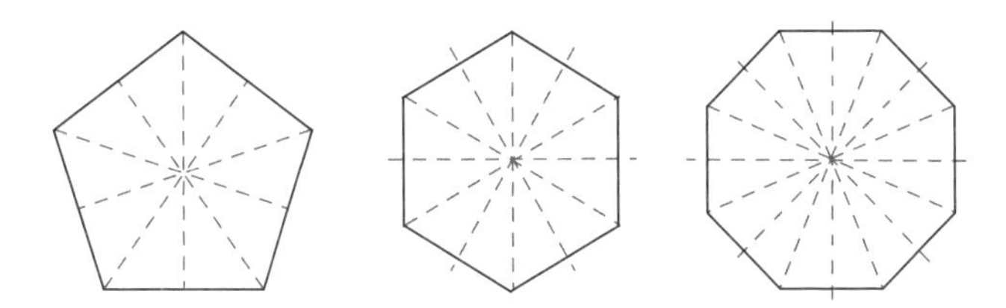

A circle has infinite lines of symmetry!

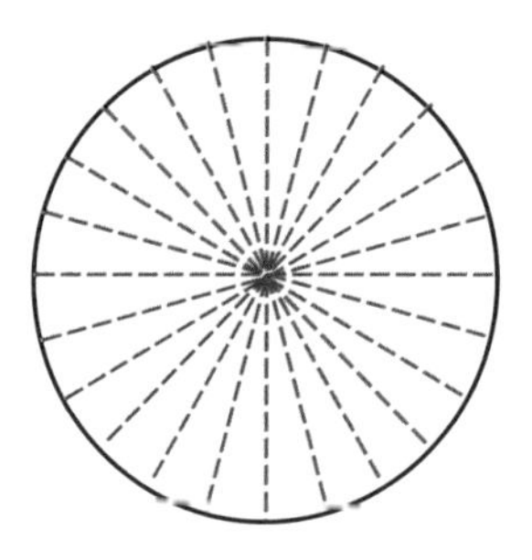

ROTATIONAL SYMMETRY

A figure has **rotational symmetry** if a rotation (other than 0° or 360°) produces the same image overlapping itself.

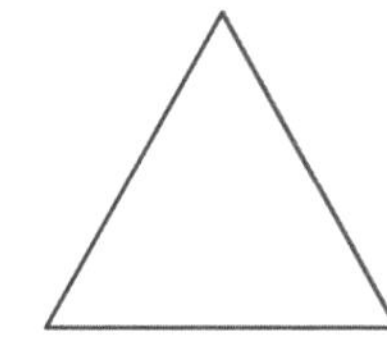
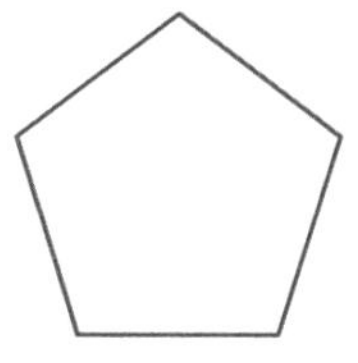

Examples of Figures with Rotational Symmetry

The **order** of rotational symmetry for a figure is the number of rotations for which the figure has symmetry.

For example, if a figure has order 2 rotational symmetry, then it looks the same at 180° and 360° rotations.

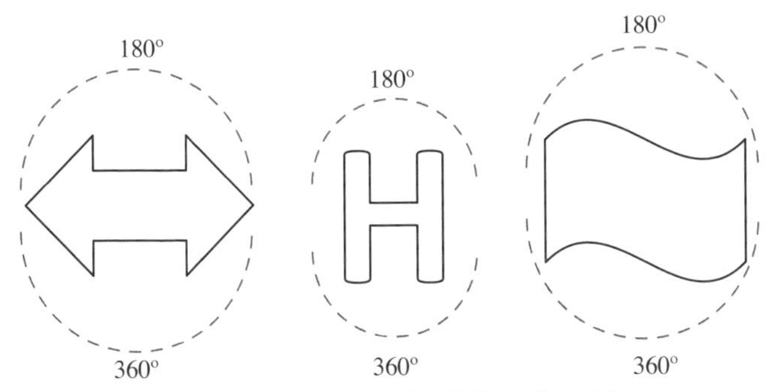

Examples of Figures with Order 2 Rotational Symmetry

There's really no such thing as "order 1" symmetry—that would just mean that something looks the same when you turn it completely around, and that's true for every figure.

If a figure has order 3 rotational symmetry, then it looks the same at 120°, 240°, and 360°.

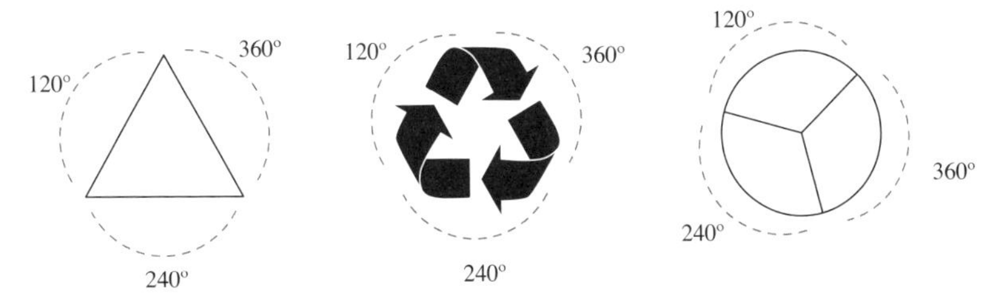

Examples of Figures with Order 3 Rotational Symmetry

And so on. The order of symmetry always divides the full rotation evenly.

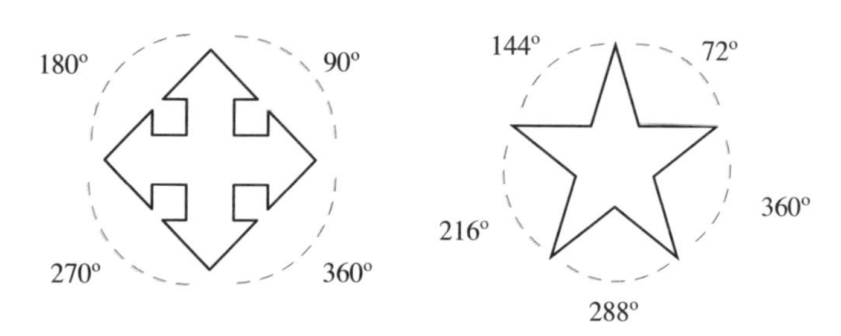

Order 4 (left) and Order 5 (right) rotational symmetry.

Any **regular polygon** has the same order of rotational symmetry as the number of its sides. For example, a regular pentagon has order 5 symmetry, a regular hexagon has order 6 symmetry, and so on.

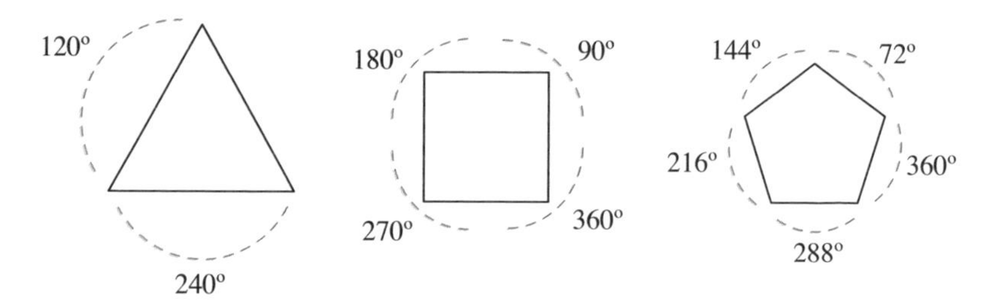

A circle has infinite rotational symmetry!

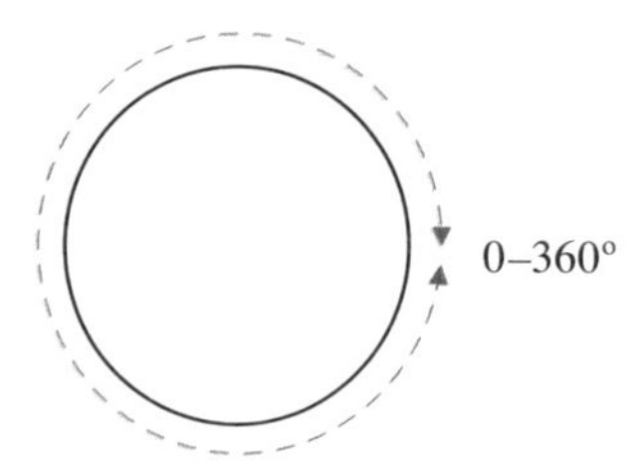

DILATIONS

In this lesson, we will review another type of image transformation—**dilation.** A dilated image is the same shape as the pre-image, but is a different size. In other words, dilation stretches or shrinks the original figure. Additionally, the image and pre-image are **similar** but not **congruent.**

In order to perform a dilation of a figure, we need to know the **scale factor** and the **center of dilation.** The scale factor is the ratio of the corresponding segment lengths of the two figures, and the center of dilation is the point of reference used to orient the figure. Additionally, if you pick any point in the dilated image, you'll be able to draw a straight line through that point, its corresponding pre-image point, and the center of dilation.

Each point in the dilated image is **collinear** with its corresponding pre-image point and the center of dilation.

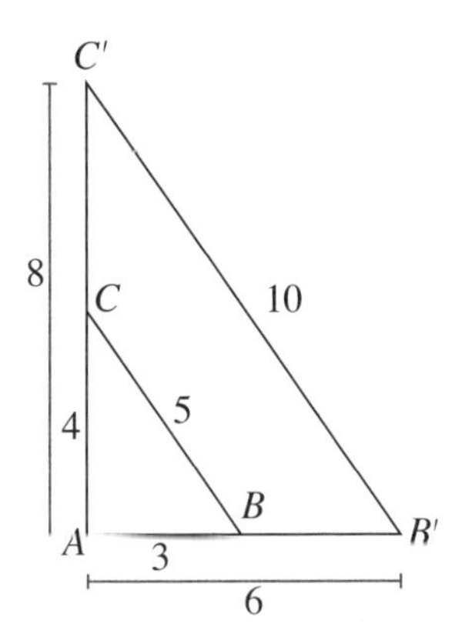

In the figure above, triangle *ABC* is dilated with a scale factor of 2 and the center of dilation is at vertex *A*.

Since the scale factor is 2, that means that the ratio of corresponding segment lengths is 1:2. In other words, the larger triangle has lengths two times greater than the smaller triangle. Therefore, the larger triangle has side lengths of 6, 8, and 10—corresponding with the smaller side lengths of 3, 4, and 5, respectively.

The center of dilation is how we position the vertices of the dilated figure. For each point in the image, the center of dilation is collinear with the image point as well as its corresponding pre-image point. In the preceding example, the center of dilation was given as vertex *A*. We have two sets of collinear vertices: points *A*, *B*, and *B'* are collinear, and points *A*, *C*, and *C'* are also collinear.

If a figure has more vertices, we'll most likely use the ratio of some of the **diagonals** for reference, rather than using only the side lengths.

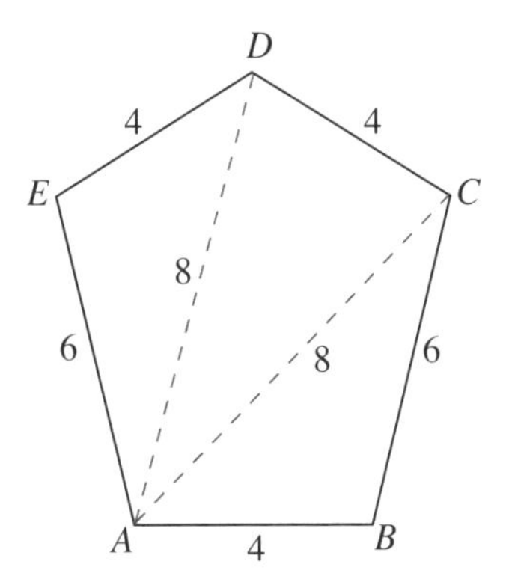

In this example, we will construct a dilated version of pentagon *ABCDE*, with scale factor of 1.5 and a center of dilation at vertex *A*.

If the scale factor is 1.5, then each segment in the dilated figure will be 1.5 times the length of the corresponding segment in the pre-image. Therefore, the segments will have the following lengths:

Pre-Image (length)	Image (length)
AB (4)	*AB'* (6)
BC (6)	*B'C'* (9)
CD (4)	*C'D'* (6)
DE (4)	*D'E'* (6)
EA (6)	*E'A* (9)
AC (8)	*AC'* (12)
AD (8)	*AD'* (12)

We're halfway there! But, how do we position the vertices in the dilated image? Recall that each point in the dilated image must be collinear with its corresponding pre-image point, as well as the center of dilation (in this case, vertex *A*). So next, we'll just need to extend each of the sides and diagonals that connect to vertex *A*.

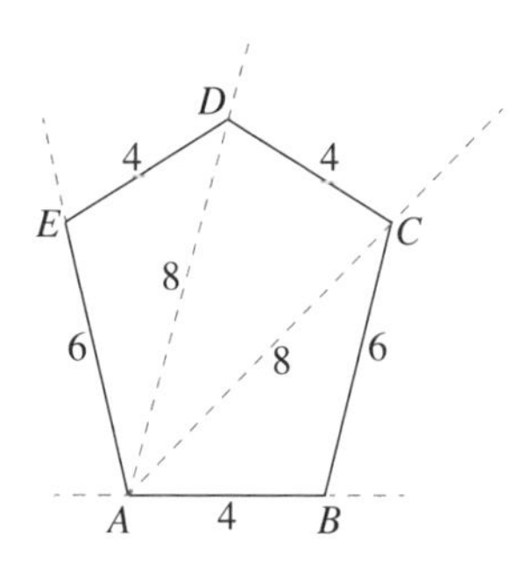

You would then measure each segment so that it matches the lengths shown in the table above. Since this is an example exercise, a scale version of the image is shown here:

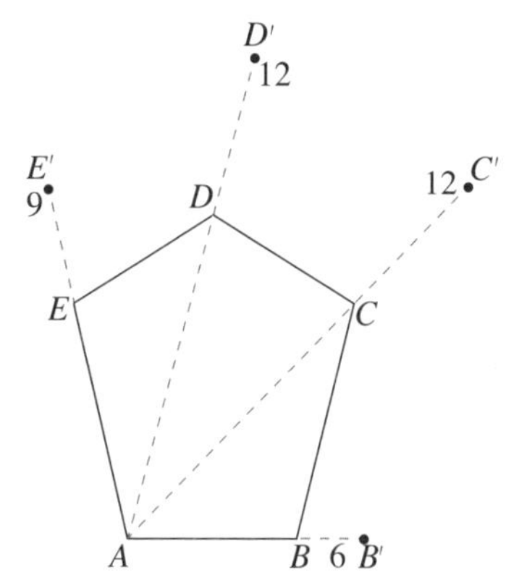

Finally, connect the new vertices to form a pentagon.

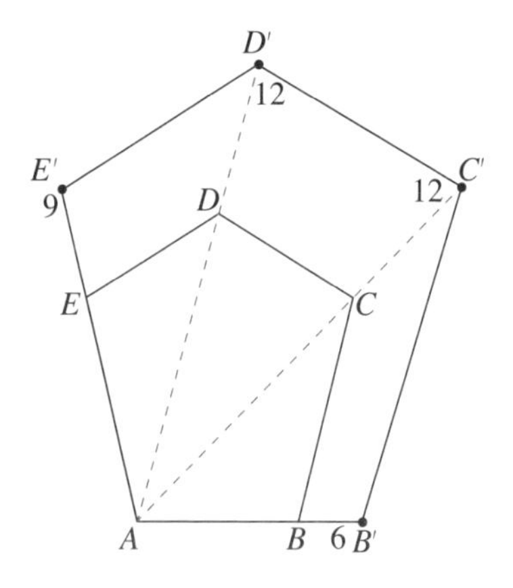

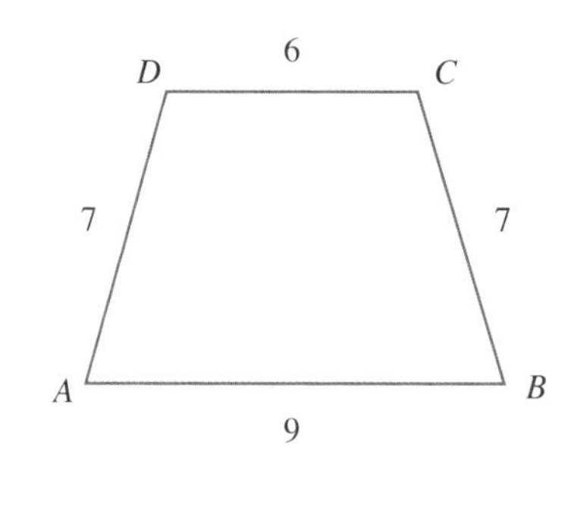

Next, we will construct a dilated version of quadrilateral *ABCD*, with scale factor 2 and center of dilation *E*.

This example is different, because the center of dilation is not one of the vertices. Thus, the given side lengths don't help us all that much, because we need to position the dilated figure with respect to point *E*. Nevertheless, the process is quite similar to the previous exercise, with one important change: you're going to measure the **distance** from each vertex to point *E*, and apply the scale ratio to those measurements.

First, draw a line from point *E* through each vertex. Then, measure the distance formed by each of these pairs of points (*EA*, *EB*, *EC*, and *ED*). Since this is an example exercise, we'll provide the "given" distances as follows:

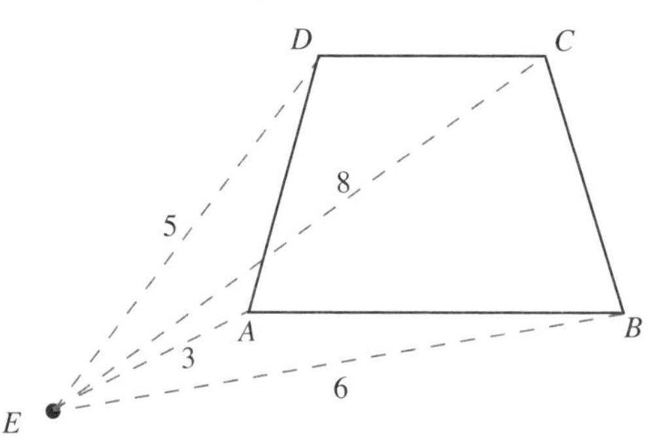

Next, extend these segments (*EA*, *EB*, *EC*, and *ED*) to two times their length. Remember to start each measurement from point *E*. Label the new endpoints *A'*, *B'*, *C'*, and *D'*.

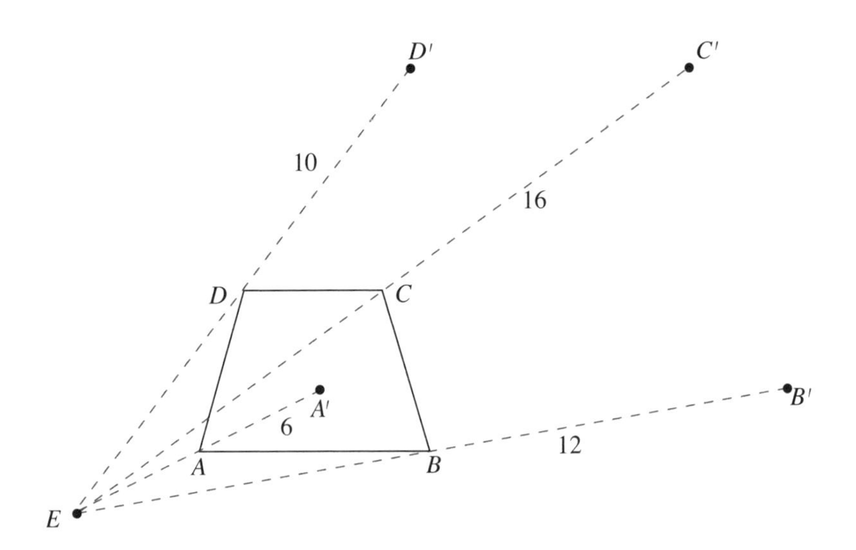

Finally, connect the four new points to form a quadrilateral.

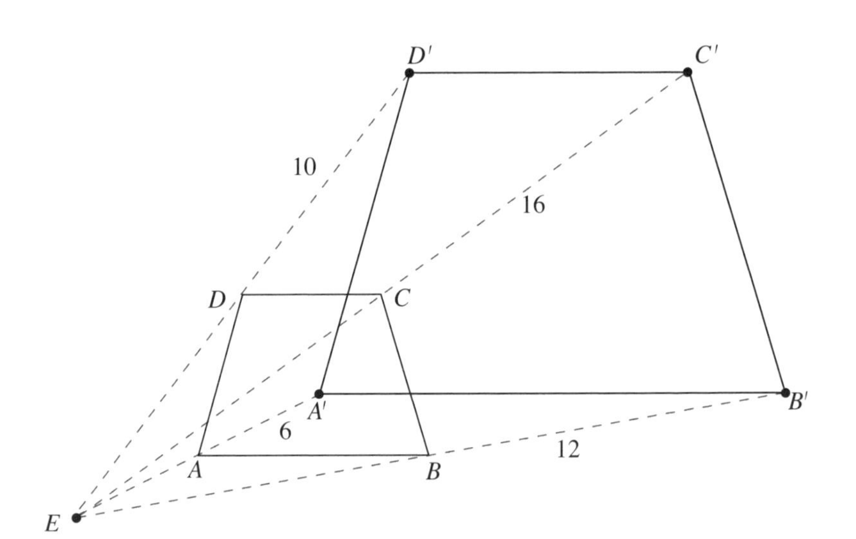

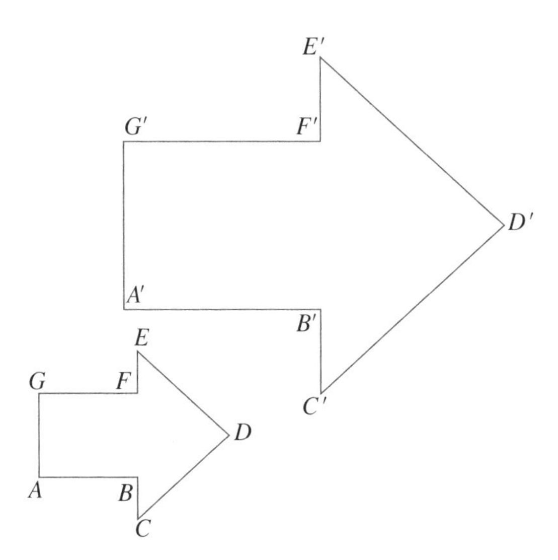

Figure *ABCDEFG* above is dilated with a scale factor of 2. In this example, we will find the center of dilation.

In this example, you're asked to find the center of dilation. This process is actually quite simple—it just involves drawing some straight lines, and you won't need to measure.

First, choose a pair of corresponding points, for instance, *A* and *A*′. Draw a long, straight line through these points. (Hint: the center of dilation is sort of to the left of point *A*, so extend the line plenty in that direction.)

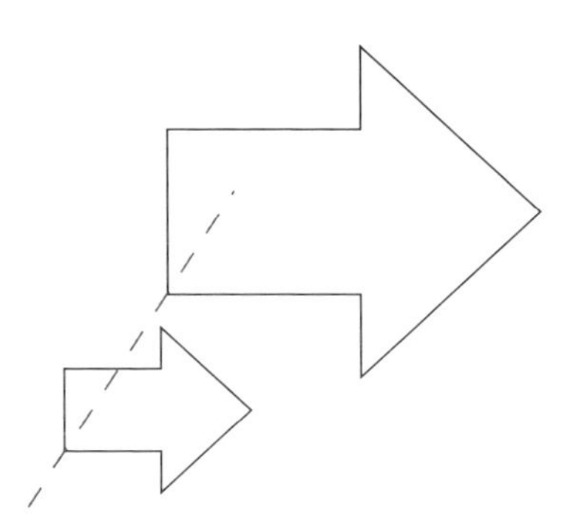

Do the same thing with a different pair of corresponding points, for instance, *B* and *B'*. Draw a long, straight line through these points.

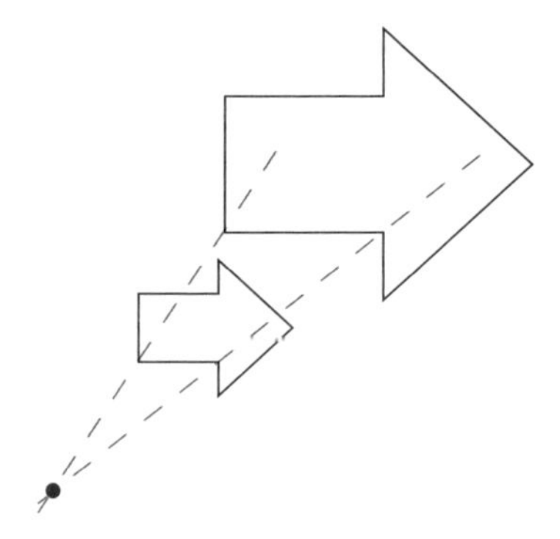

Where these two lines intersect is the center of dilation.

You only need to draw two lines, but it doesn't hurt to add the remaining lines to check your work.

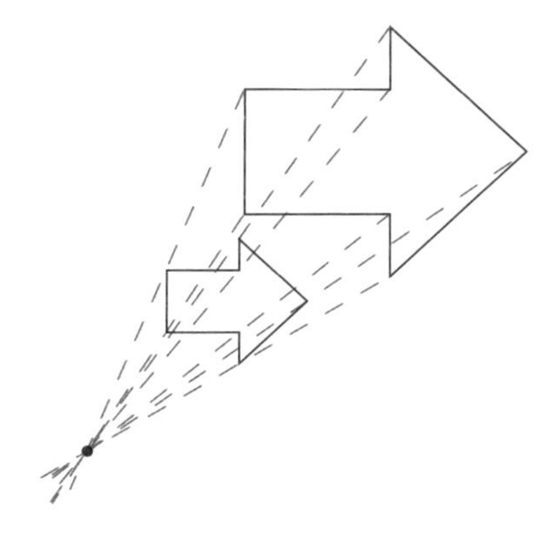

DILATIONS IN THE COORDINATE PLANE

If the center of dilation is the **origin** (point 0, 0), you'll apply the scale factor multiple to each reference point in the figure. For example, if the scale factor is 4, then point (2, 3) would be dilated to (8, 12).

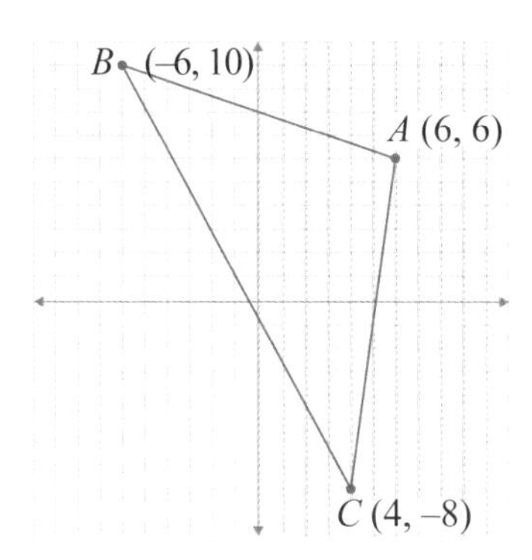

Here, we will construct a dilated version of triangle *ABC* with a scale factor of $\frac{1}{2}$ and the center of dilation at (0, 0).

When the center of dilation is at the origin, all you need to do is multiply each coordinate by the scale factor $\left(\frac{1}{2}\right)$. Complete this process for each of the three vertices, and find the image coordinates as follows:

Pre-Image	Image
A (6, 6)	*A'* (3, 3)
B (–6, 10)	*B'* (–3, 5)
C (4, –8)	*C'* (2, –4)

Finally, plot the three points and then connect the three vertices to form a triangle.

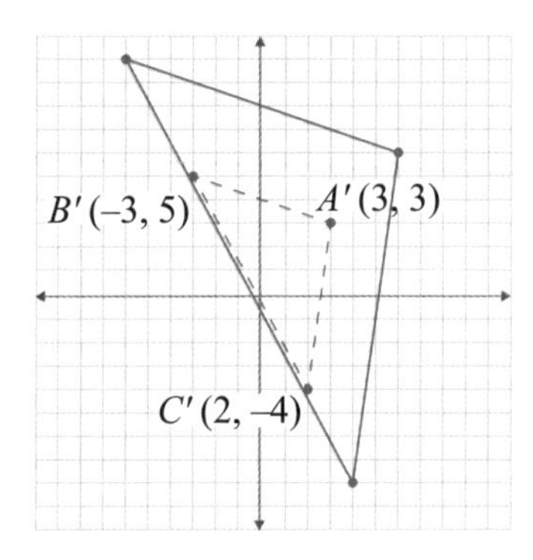

If the center of dilation is *not* at the origin, then you'll use the same process that we reviewed earlier in this chapter—applying the scale factor multiple to the *distance from the center of dilation*. This is a little easier than it might sound!

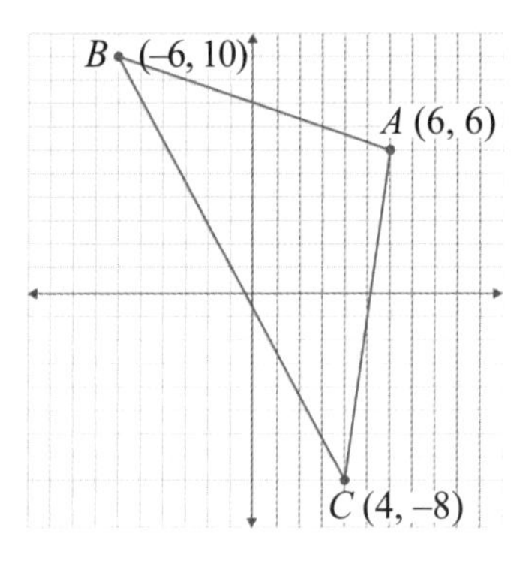

In this example, we will construct a dilated version of triangle *ABC* with a scale factor of $\frac{1}{2}$ and the center of dilation at (–2, –4).

Let's begin with vertex *A* (6, 6). We can apply the same process as we did with non-coordinate dilation, extending lines from the center of dilation and measuring the distance.

Draw a line from the center of dilation and through coordinate *A*. Of course, the distance wasn't given here, so we'll have to find it. The good news is that we won't need to use Distance Formula or anything overly complicated—we'll just focus on the horizontal and vertical difference between the points. By counting, we can see that coordinate *A* is **8** horizontal units and **10** vertical units from the center of dilation.

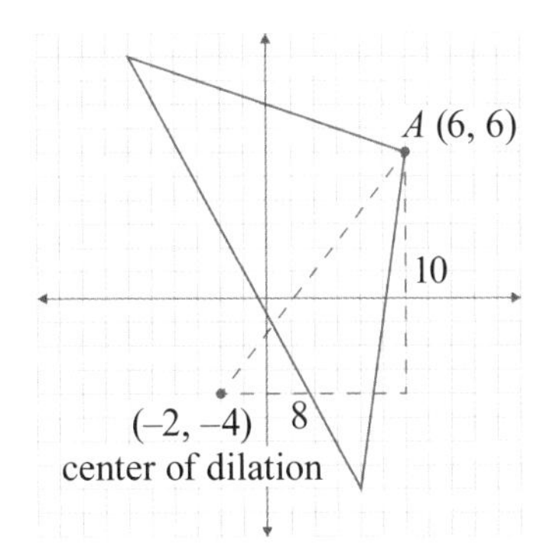

The scale factor for this exercise is $\frac{1}{2}$. That means that the dilated image of *A* (labeled *A′* in the figure below) should be **4** horizontal units and **5** vertical units (in the same direction) from the center of dilation. In other words, the distance between *A′* and (–2, –4) is $\frac{1}{2}$ the distance between *A* and (–2, –4). The coordinate *A′* is located at (2, 1).

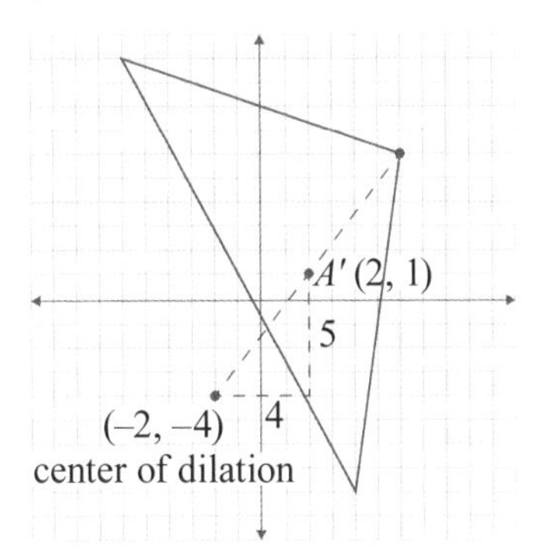

Repeat this process for coordinates *B* and *C*.

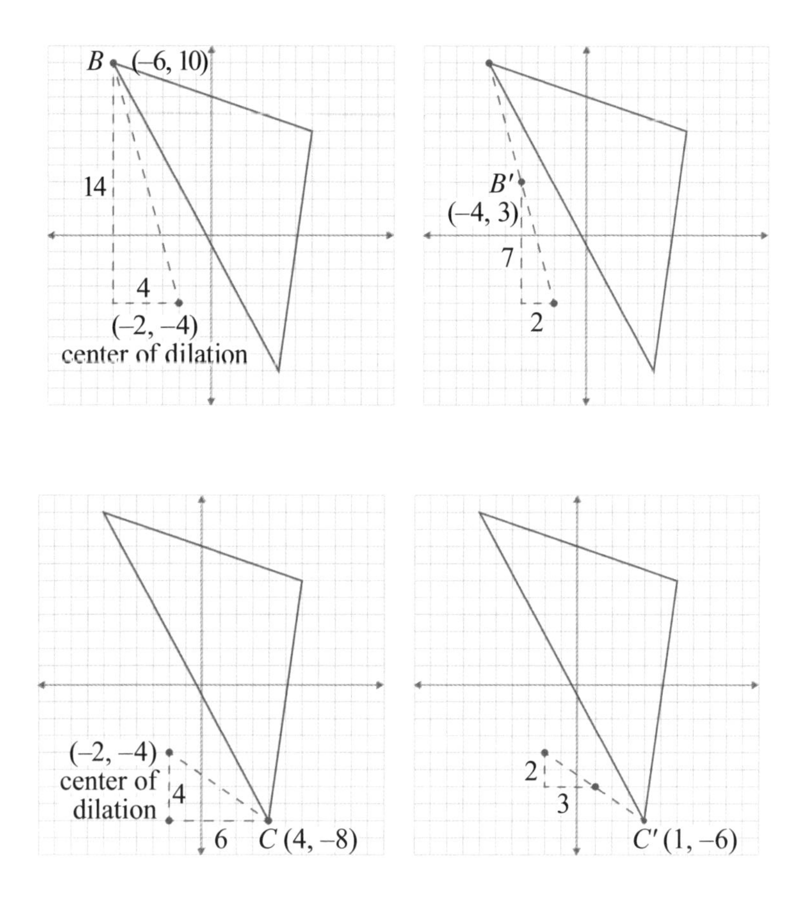

Finally, connect the vertices to form a triangle.

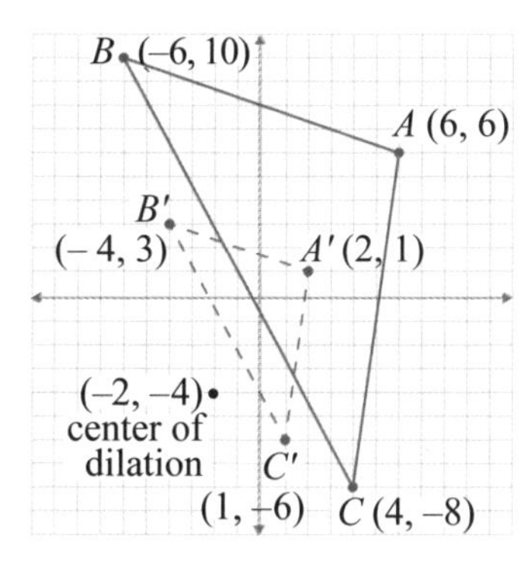

CHAPTER

TRIGONOMETRY

Trigonometry is the study of triangles. It's actually so important that there is an entire branch of mathematics devoted to this topic. The most common ways that you'll use trigonometry in high school include solving for unknown side lengths and/or angles in triangles. In the real world, trigonometry has many applications in fields such as physics, engineering, astronomy, and even music.

CHAPTER CONTENTS

TRIGONOMETRIC RATIOS

Almost everything we know about trigonometry can be derived from relationships found in right triangles. In fact, some basics that you already know—like the Pythagorean Theorem—play a fundamental role. Trigonometry allows us to use the known ratios of a triangle to solve for unknown information, like side lengths.

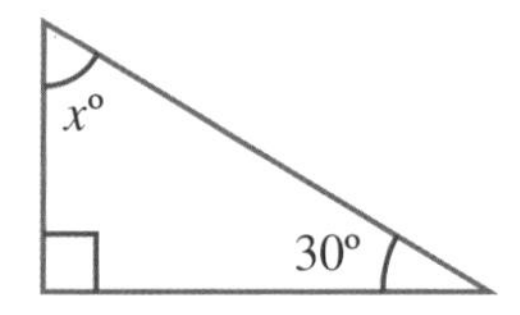

Looking at the triangle above, you know that one angle is 90° and one angle is 30°. What else do you know? Since the angles in a triangle always add up to 180°, you'd also be able to find *x*. 180° – (90° + 30°) = 60°, so *x* would be 60°. You have all three angles for this triangle, then, which also means you know its proportions. That's the fundamental theorem of trigonometry—if you know the angles of a triangle, you know the ratios of the sides. That is, you might not know the actual values of the individual sides, but you know the relationships between them.

Based on what we know of triangle proportions, trigonometry has three basic functions to express these relationships.

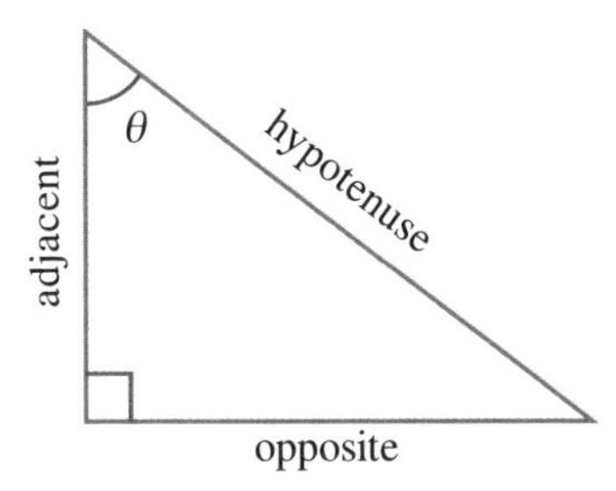

Here, we'll define some terms. Then, you'll see some examples.

θ (called "theta")—a commonly used variable for angles. (Think of it like an x.)

Opposite and *adjacent*—the two **legs** of a triangle.

The *opposite* side—the leg that's opposite from (i.e., not touching) the angle θ.

The *adjacent* side—the leg that's adjacent to (i.e., touching) the angle θ.

The *hypotenuse*—the longest side of a right triangle. It's always opposite from the 90° angle.

Sine

The sine of an angle is the ratio of its opposite side to the hypotenuse.

The sine function of an angle θ is abbreviated as $\sin \theta$.

Cosine

The cosine of an angle is the ratio of its adjacent side to the hypotenuse.

The cosine function of an angle θ is abbreviated as $\cos \theta$.

Tangent

The tangent of an angle is the ratio of its opposite side to its adjacent side.

The tangent function of an angle θ is abbreviated as $\tan \theta$.

A mnemonic for these functions is SOHCAHTOA (pronounce it like "so-ca-toe-a").

$$S = \frac{O}{H}$$

$$C = \frac{A}{H}$$

$$T = \frac{O}{A}$$

SOHCAHTOA

sin = $\frac{\textbf{o}\text{pposite}}{\textbf{h}\text{ypotenuse}}$ **c**os = $\frac{\textbf{a}\text{djacent}}{\textbf{h}\text{ypotenuse}}$ **t**an = $\frac{\textbf{o}\text{pposite}}{\textbf{a}\text{djacent}}$

Let's take another look at the 30°-60°-90° triangle. Remember that this is a "special" right triangle, whose proportions you may have memorized previously.

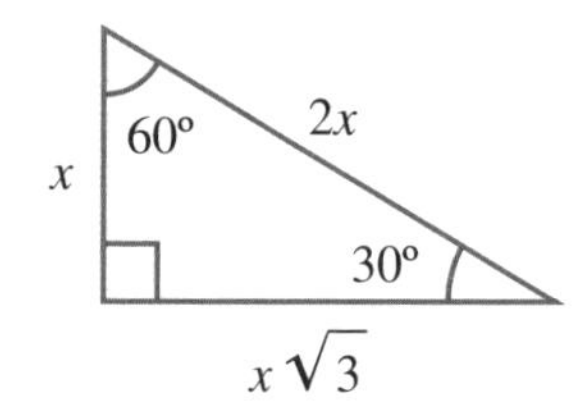

For example, if the short side is 2, then the triangle would have the following side lengths.

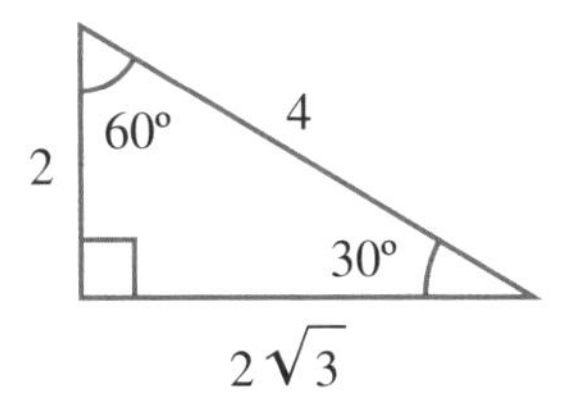

REMEMBER...

"Opposite" and "adjacent" are relative to the angle you're working with. The "adjacent" of 30° is the same as the "opposite" of 60°, and vice versa.

Here are the trigonometric ratios for these angles.

	30°	60°
Sine *(opposite/ hypotenuse)*	$=\frac{2}{4}=\frac{1}{2}=0.5$	$=\frac{2\sqrt{3}}{4}=\frac{\sqrt{3}}{2}\approx 0.866$
Cosine *(adjacent/ hypotenuse)*	$=\frac{2\sqrt{3}}{4}=\frac{\sqrt{3}}{2}\approx 0.866$	$=\frac{2}{4}=\frac{1}{2}=0.5$
Tangent *(opposite/ adjacent)*	$=\frac{2}{2\sqrt{3}}=\frac{1}{\sqrt{3}}$ (or, $\frac{\sqrt{3}}{3}$) ≈ 0.577	$=\frac{2\sqrt{3}}{2}=\sqrt{3}\approx 1.732$

Fractions in simplified form do not have irrational numbers in the denominator. To simplify, multiply the irrational number on the top and bottom of the fraction.

Example:

$$\frac{1}{\sqrt{3}} = \frac{1 \times \sqrt{3}}{\sqrt{3} \times \sqrt{3}} = \frac{\sqrt{3}}{3}$$

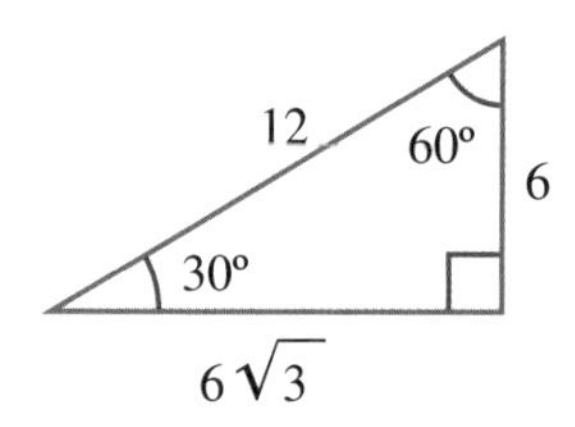

This triangle is larger than the one in the previous example. However, it is still a 30°-60°-90° triangle. In other words, the two triangles are similar—they have the same angles, even though they are different sizes.

Using these side lengths, we can see that the trigonometric ratios are still the same.

$$\sin 30° = \frac{6}{12} = \frac{1}{2}$$

$$\cos 30° = \frac{6\sqrt{3}}{12} = \frac{\sqrt{3}}{2}$$

$$\tan 30° = \frac{6}{6\sqrt{3}} = \frac{1}{\sqrt{3}}$$

Recall the AAA Similarity Postulate from Chapter 5

Therefore, since similar triangles have the same angles, they also have the same trigonometric relationships.

Try practicing the trigonometric ratios with a 3-4-5 triangle or a 45°-45°-90° triangle!

Each angle has given values for sine, cosine, and tangent.

COMPLEMENTARY ANGLES

In geometry, two angles are **supplementary** to each other if their sum equals 180°. For example, 45° is supplementary to 135°.

When two angles have a sum of 90°, they are known as **complementary** angles. For example, 30° is complementary to 60°. In a right triangle, the two acute angles are always complementary.

The concept of complementary angles is very important in trigonometry. You may have noticed in the examples of the right triangles in the previous section, that the complementary angles have the same sin and cos values—except that they're switched. That is, the sine of one angle is equal to the cosine of its complement. For example, the sine of 30° is equal to the cosine of 60°. (Both are equal to 0.5.)

Additionally, the tangents of complementary angles are **reciprocals** of each other. In other words, the tangent of one angle is equal to the reciprocal of the tangent of its complement. For example, the tangent of 60° is $\sqrt{3}$, while the tangent of 30° is $\frac{1}{\sqrt{3}}$.

Complementary Angles

Complementary angles have a sum of 90°.

The sine of x° is equal to the cosine of 90 – x°.

The cosine of x° is equal to the sine of 90 – x°.

The tangent of x° is the reciprocal of the tangent of 90 – x°.

YOUR CALCULATOR

Scientific and graphing calculators have sin, cos, and tan built in! When given an angle, the calculator can tell you the value of sin, cos, or tan for that angle.

Want to try it? With your calculator in degree mode, enter sin(30°). The calculator should show $\frac{1}{2}$, or 0.5.

> It is not typically practical to solve inverse functions without a calculator. The inverse function of some "special" right triangles (such as 30°-60°-90° or 45°-45°-90°) can be memorized. One other way to solve inverse functions is to observe a graph of sin, cos, or tan values.

You can try the different functions for 30° and 60°, and compare the results with the values in the table shown previously.

Your calculator can also do the *inverse* of these functions, which takes the *ratio* and solves for the *angle.* For example, the inverse of sin30° is just 30°. On your calculator, the inverse of sin might look like $\sin^{-1}$, or "arcsin."

If you enter $\sin^{-1}(0.5)$, in degree mode, your calculator should return 30°.

Suppose a contractor builds a ramp to reach a loading dock that is 10 feet high. The ramp measures 26 feet along its slanted surface. We can use an inverse trigonometric function to find the angle of incline for the ramp.

The ramp forms a right triangle, in which the *opposite* side is 10 feet and the *hypotenuse* is 26 feet. Since those are the two given sides, the most straightforward way to solve this problem is to calculate $\sin^{-1}\left(\frac{10}{26}\right)$, or simplify as $\sin^{-1}\left(\frac{5}{13}\right)$. Use your calculator.

$$\sin^{-1}\left(\frac{5}{13}\right) \approx 22.62°$$

Note that you can also solve for the unknown side of this triangle, using the Pythagorean Theorem.

$$10^2 + b^2 = 26^2$$
$$100 + b^2 = 676$$
$$b^2 = 676 - 100$$
$$b^2 = 576$$
$$b = 24$$

Knowing all three sides of the triangle, you can now use the other two inverse identities to solve for the same angle.

$$\sin^{-1}\left(\frac{5}{13}\right) \approx 22.62°$$

$$\cos^{-1}\left(\frac{12}{13}\right) \approx 22.62°$$

$$\tan^{-1}\left(\frac{5}{12}\right) \approx 22.62°$$

Did we *need* to do that here? Of course not. However, it's helpful to know your options for more difficult problems. Some exercises in school, or on standardized tests, may challenge you to use specific functions, instead of the most straightforward one.

OTHER TRIGONOMETRIC IDENTITIES

RECIPROCAL IDENTITIES

The reciprocals of sine, cosine, and tangent are, respectively, **cosecant** (abbreviated csc), **secant** (abbreviated sec), and **cotangent** (abbreviated cot). This builds on the relationships of sin, cos, and tan, and can come in handy with more advanced problems. For now, just work on memorizing these relationships.

$$\csc = \frac{\text{hypotenuse}}{\text{opposite}} = \frac{1}{\sin}$$

$$\sec = \frac{\text{hypotenuse}}{\text{adjacent}} = \frac{1}{\cos}$$

$$\cot = \frac{\text{adjacent}}{\text{opposite}} = \frac{1}{\tan}$$

Observe the ratios in the triangle below.

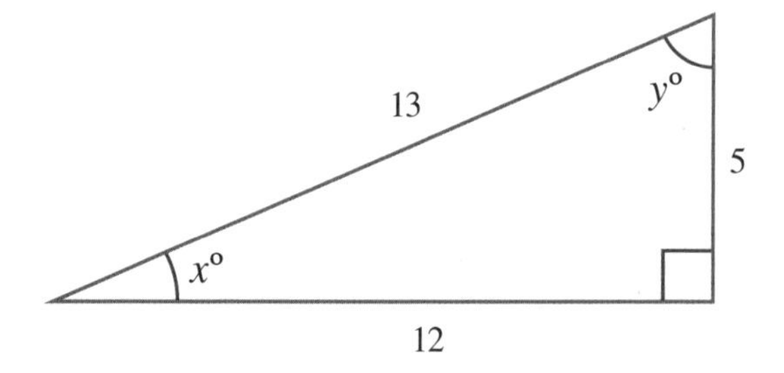

	x°	y°
Sine *(opposite/ hypotenuse)*	$=\frac{5}{13}$ ≈ 0.385	$=\frac{12}{13}$ ≈ 0.923
Cosine *(adjacent/ hypotenuse)*	$=\frac{12}{13}$ ≈ 0.923	$=\frac{5}{13}$ ≈ 0.385
Tangent *(opposite/ adjacent)*	$=\frac{5}{12}$ ≈ 0.417	$=\frac{12}{5}$ $= 2.4$
Cosecant *(hypotenuse/ opposite)*	$=\frac{13}{5}$ $= 2.6$	$=\frac{13}{12}$ ≈ 1.083
Secant *(hypotenuse/ adjacent)*	$=\frac{13}{12}$ ≈ 1.083	$=\frac{13}{5}$ ≈ 2.6
Cotangent *(adjacent/ opposite)*	$=\frac{12}{5}$ $= 2.4$	$=\frac{5}{12}$ ≈ 0.417

THE PYTHAGOREAN IDENTITY

For all angles, the value of sin and cos is always between −1 and 1 (inclusive).

Tan does not have this limit.

Remember: sine of $x°$ is written as $\sin(x°)$.

Theta (θ) is a variable for an angle.

The Pythagorean Identity is based on—you guessed it—the Pythagorean Theorem. Let's see how this is derived.

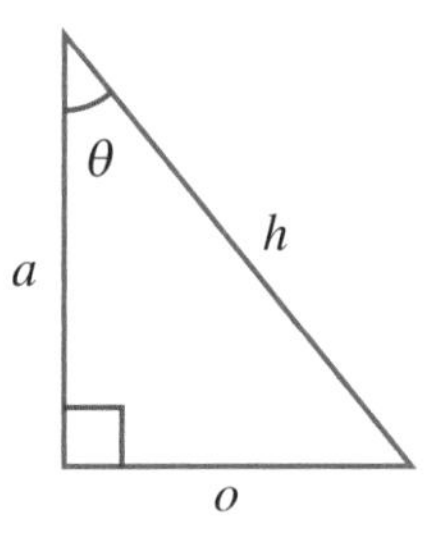

$$\sin^2(\theta) + \cos^2(\theta) = 1$$

See this in action with a 3-4-5 triangle:

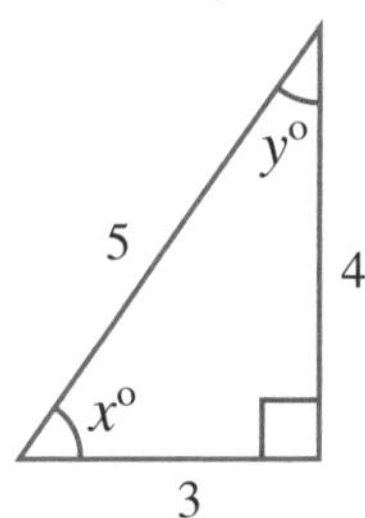

An **identity** is an equation or formula that's true for all values.

$$\sin(x) = \frac{3}{5}$$

$$\cos(x) = \frac{4}{5}$$

$$\sin^2(x) = \left(\frac{4}{5}\right)^2 = \frac{16}{25}$$

$$\cos^2(x) = \left(\frac{3}{5}\right)^2 = \frac{9}{25}$$

$$\sin^2(\theta) + \cos^2(\theta) = \frac{16}{25} + \frac{9}{25} = \frac{25}{25} = 1$$

Here's another way to think about this identity.

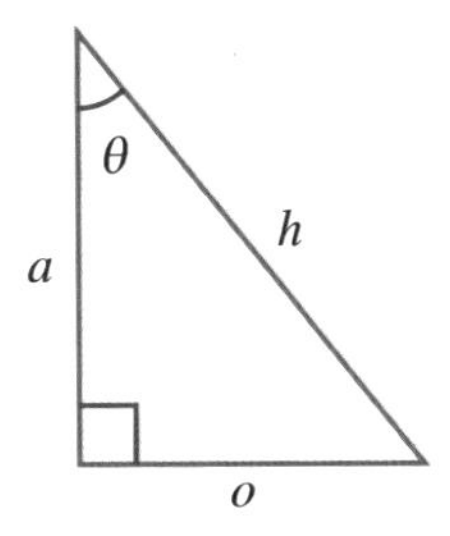

$o^2 + a^2 = h^2$	Start with the Pythagorean Theorem. Since this is a right triangle, we know the theorem will be satisfied.
$\frac{o^2}{h^2} + \frac{a^2}{h^2} = \frac{h^2}{h^2}$ $\frac{o^2}{h^2} + \frac{a^2}{h^2} = 1$	Divide both sides by h^2.
$\sin^2(\theta) + \cos^2(\theta) = 1$	Observe that $\frac{o^2}{h^2} = \left(\frac{o}{h}\right)^2 = \sin^2(\theta)$. Observe that $\frac{a^2}{h^2} = \left(\frac{a}{h}\right)^2 = \cos^2(\theta)$.

From this identity, we can derive all of the following:

$$\sin^2(\theta) + \cos^2(\theta) = 1$$

$$1 - \sin^2(\theta) = \cos^2(\theta)$$

$$1 - \cos^2(\theta) = \sin^2(\theta)$$

$$\tan^2(\theta) + 1 = \sec^2(\theta)$$

$$\cot^2(\theta) + 1 = \csc^2(\theta)$$

...and more!

SIN/COS

Another very useful identity is the following:

$$\frac{\sin(\theta)}{\cos(\theta)} = \tan(\theta)$$

Let's see how this identity is derived.

$\frac{\sin(\theta)}{\cos(\theta)}$	$\sin(\theta) = \frac{o}{h}$ $\cos(\theta) = \frac{a}{h}$
$= \frac{\frac{o}{h}}{\frac{a}{h}} = \frac{o}{h} \times \frac{h}{a} = \frac{oh}{ha} = \frac{o}{h} = \tan(\theta)$	To divide, multiply by the reciprocal. Cancel *h* from the numerator and denominator. $\tan(\theta) = \frac{o}{a}$

Let's see this identity with a 5-12-13 triangle.

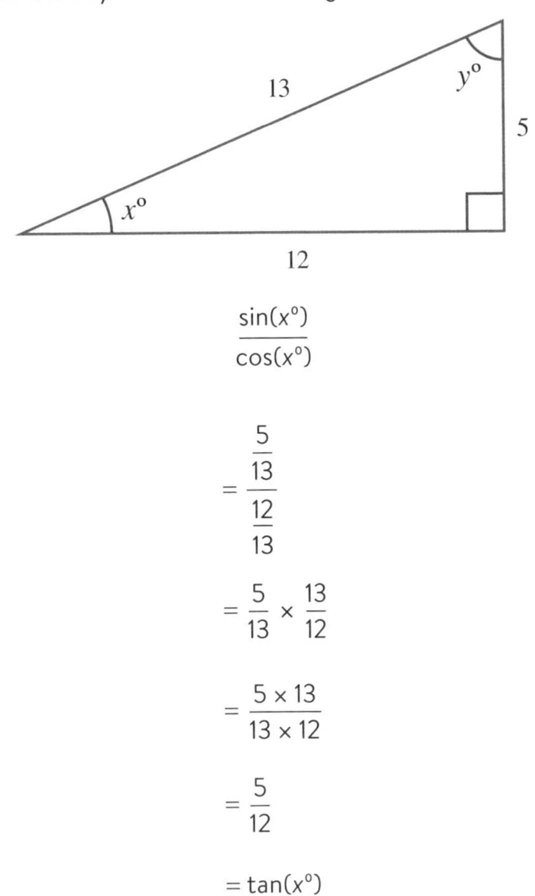

$$\frac{\sin(x^\circ)}{\cos(x^\circ)}$$

$$= \frac{\frac{5}{13}}{\frac{12}{13}}$$

$$= \frac{5}{13} \times \frac{13}{12}$$

$$= \frac{5 \times 13}{13 \times 12}$$

$$= \frac{5}{12}$$

$$= \tan(x^\circ)$$

From this identity, we can derive the following:

$$\frac{\sin(\theta)}{\cos(\theta)} = \tan(\theta)$$

$$\frac{\cos(\theta)}{\sin(\theta)} = \cot(\theta)$$

...and more!

TRIGONOMETRY WITH NON-RIGHT TRIANGLES

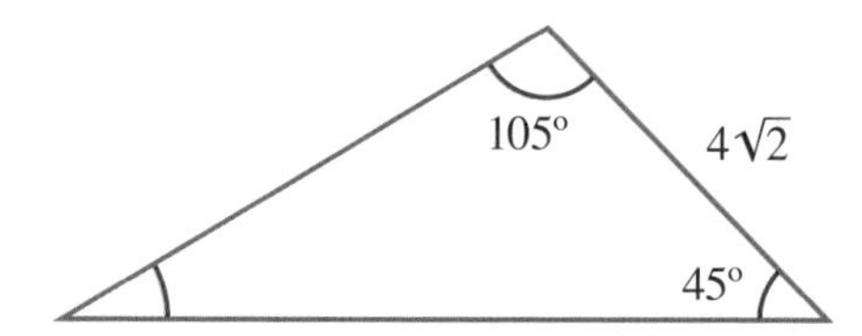

Suppose you need to find the area of the triangle above. First, you'll need to find the lengths of the height and base. Draw a line representing the height. Label the unknown sides as *a*, *b*, and *c*. The height must always be perpendicular to the base, so you know it forms a right angle. Therefore, the smaller triangle with 45° and 90° must be a 45°-45°-90° triangle.

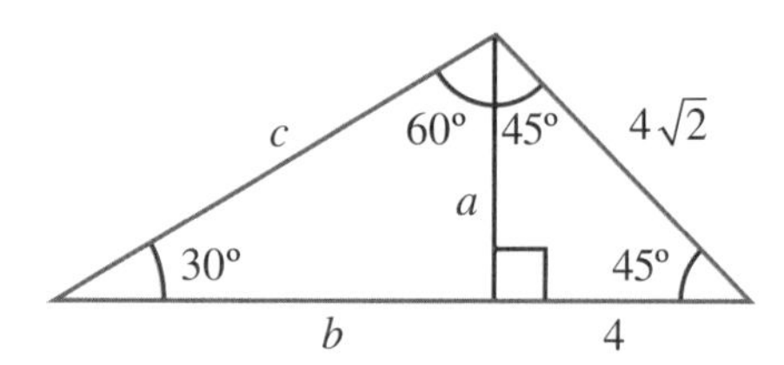

You can use the given side length of $4\sqrt{2}$ to find more information. $4\sqrt{2}$ is the hypotenuse of our 45°-45°-90° triangle. Therefore, the two legs of this smaller triangle are each equal to 4.

To prove this, recall the "special" right triangle 45°-45°-90° has side lengths *x*, *x*, and $x\sqrt{2}$.

Or, you can use the Pythagorean Theorem: $a^2 + a^2 = \left(4\sqrt{2}\right)^2$.
(The 45°-45°-90° triangle is isosceles.)

Or, you can use $\sin(45°) = \frac{a}{4\sqrt{2}}$.

That's a lot of options!

$$\text{height} = 4$$

Now that you know the height of this triangle, you can also find the other unknown side lengths.

If the height is 4, then b is $4\sqrt{3}$, and c must be 8.

To prove this, use the relationships of the "special" right triangle 30°-60°-90°.

Or, use sin, cos, or tan functions to solve. (For example, use $\tan(60°) = \frac{b}{4}$ and $\cos(60°) = \frac{c}{4}$).

$$\text{base} = 4\sqrt{3} + 4$$
$$\approx 10.93$$

Therefore, the area of the triangle is determined as follows:

$$A = \frac{1}{2}bh$$
$$\approx \frac{1}{2} \times 4 \times (10.93)$$
$$\approx 2 \times (10.93)$$
$$\approx 21.86$$

THE LAW OF SINES

Because of the ability to make triangles within triangles, yet another identity can be derived.

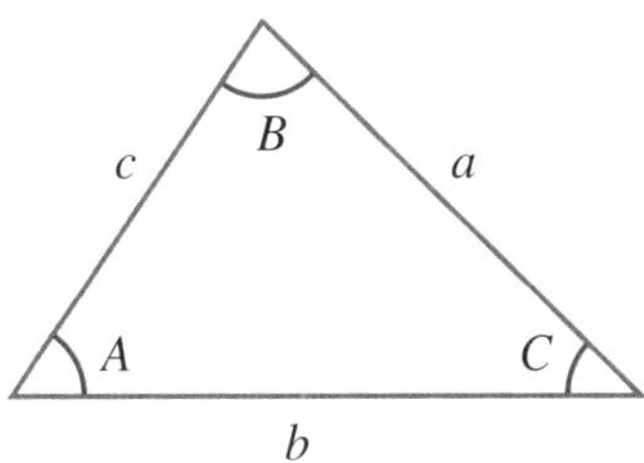

The Law of Sines

$$\frac{\sin(A)}{a} = \frac{\sin(B)}{b} = \frac{\sin(C)}{c}$$

in which side *a* is opposite angle *A*, side *b* is opposite angle *B*, and side *c* is opposite angle *C*.

The way to use the Law of Sines is to create a proportion. Each angle corresponds with its opposite side.

Let's take another look at the 105-30-45 triangle.

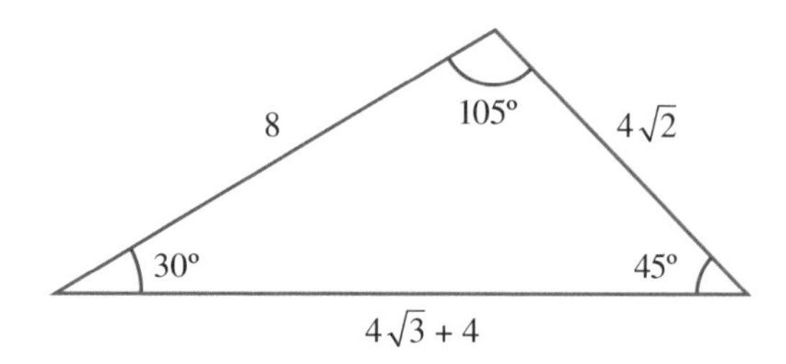

Using the Law of Sines, we have the following proportion:

$$\frac{\sin(30^\circ)}{4\sqrt{2}} = \frac{\sin(45^\circ)}{8} = \frac{\sin(105^\circ)}{4\sqrt{3}+4}$$

Use your calculator to confirm that the proportion is true:

$$\frac{\sin(30^\circ)}{4\sqrt{2}} \approx 0.0884$$

$$\frac{\sin(45^\circ)}{8} \approx 0.0884$$

$$\frac{\sin(105^\circ)}{4\sqrt{3}+4} \approx 0.0884$$

We can use the Law of Sines to solve for unknown side lengths or angles.

THE LAW OF COSINES

The Law of Cosines is based on three sides and one angle. It allows you to solve if one of those facts is unknown.

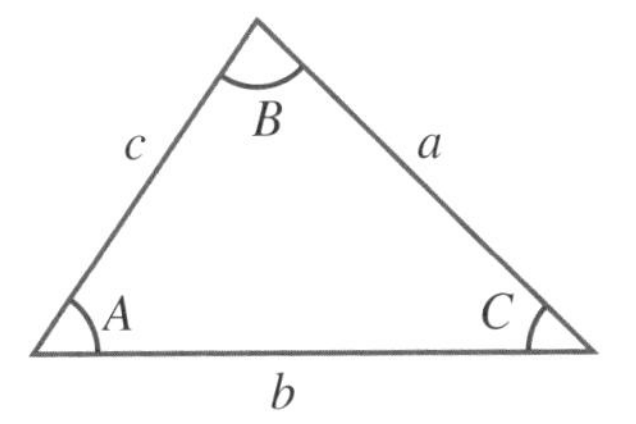

The Law of Cosines

$c^2 = a^2 + b^2 - 2ab\,(\cos(C))$

in which side *a* is opposite angle *A*, side *b* is opposite angle *B*, and side *c* is opposite angle *C*.

See it in action with a 30°-60°-90° triangle.

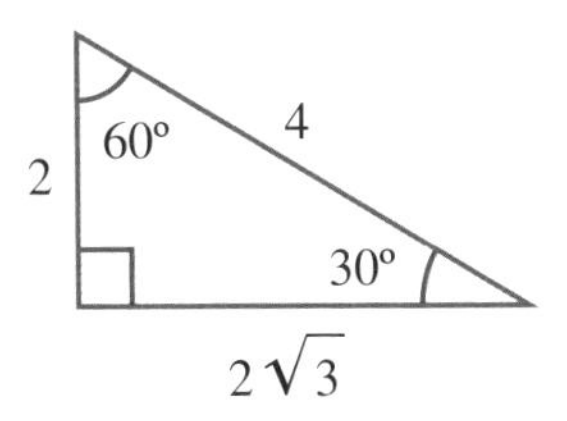

REMEMBER...

The most important thing about setting up the Law of Cosines is that the angle C is opposite side *c*.

There are a few different ways we can set up the Law of Cosines with this figure.

$$2^2 = 4^2 + \left(2\sqrt{3}\right)^2 - 2\left(4 \times 2\sqrt{3}\right)\left(\cos(30^\circ)\right)$$

$$\left(2\sqrt{3}\right)^2 = 4^2 + 2^2 - 2\left(4 \times 2\right)\left(\cos(60^\circ)\right)$$

$$4^2 = 2^2 + \left(2\sqrt{3}\right)^2 - 2\left(2 \times 2\sqrt{3}\right)\left(\cos(90^\circ)\right)$$

There are no unknowns in this example, so just simplify and use your calculator to see that the equations are true. We'll do the first one step-by-step:

$$2^2 = 4^2 + \left(2\sqrt{3}\right)^2 - 2\left(4 \times 2\sqrt{3}\right)\left(\cos(30^\circ)\right)$$

$$4 = 4^2 + \left(2\sqrt{3}\right)^2 - 2\left(4 \times 2\sqrt{3}\right)\left(\cos(30^\circ)\right)$$

$$4 = 16 + \left(2\sqrt{3}\right)^2 - 2\left(4 \times 2\sqrt{3}\right)\left(\cos(30^\circ)\right)$$

$$4 = 16 + 12 - 2\left(4 \times 2\sqrt{3}\right)\left(\cos(30^\circ)\right)$$

$$4 = 16 + 12 - 2\left(8\sqrt{3}\right)\left(\cos(30^\circ)\right)$$

$$4 = 16 + 12 - \left(16\sqrt{3}\right)\left(\cos(30^\circ)\right)$$

$$4 \approx 16 + 12 - (27.71)(\cos(30^\circ))$$

$$4 \approx 16 + 12 - (27.713)(0.866)$$

$$4 \approx 16 + 12 - (23.99)$$

$$4 \approx 28 - (23.99)$$

$$4 \approx 4.01$$

That's about right! If you rounded the irrational numbers, you can expect the answer to be inexact.

CHAPTER

THREE-DIMENSIONAL FIGURES

We live in a three-dimensional world, so why should we work only in two-dimensional geometry? It may sound intimidating, but working in three dimensions isn't all that different from working in two dimensions.

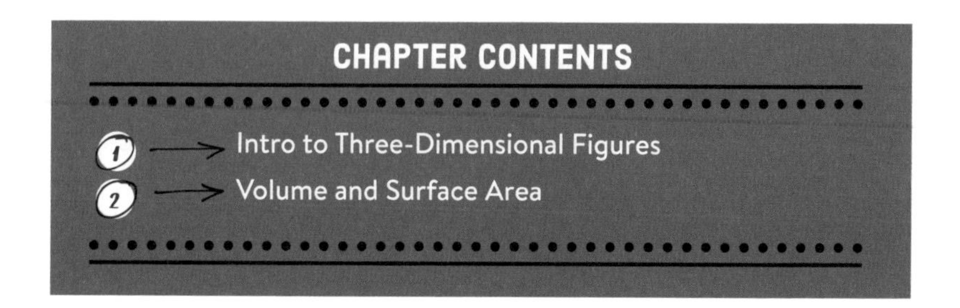

INTRO TO THREE-DIMENSIONAL FIGURES

POLYHEDRA

A **polyhedron** (plural **polyhedra**) is a three-dimensional shape for which each face is a flat surface. In other words, each face is a **polygon**. The faces intersect each other as straight line segments, and these edges intersect each other at single-point vertices.

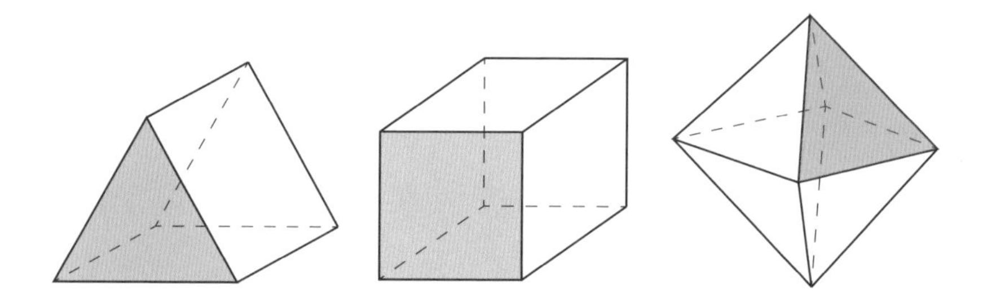

A **regular polyhedron** is one whose faces are all congruent, regular polygons. Also, in a regular polyhedron, the same number of faces meet at each vertex. There are only five types of regular convex polyhedra, and they have special names (see the figures and descriptions on the next page).

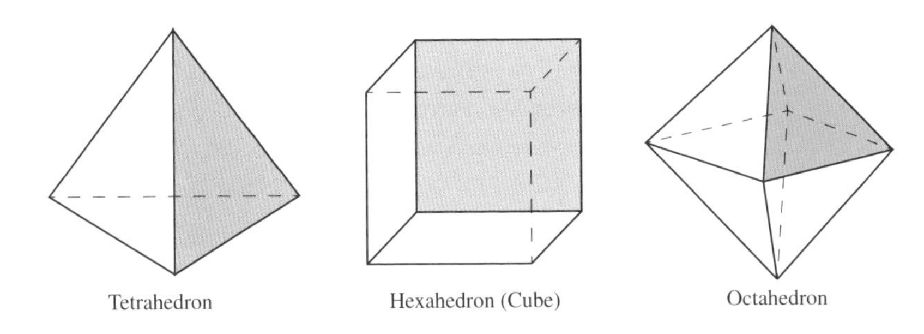

Tetrahedron   Hexahedron (Cube)   Octahedron

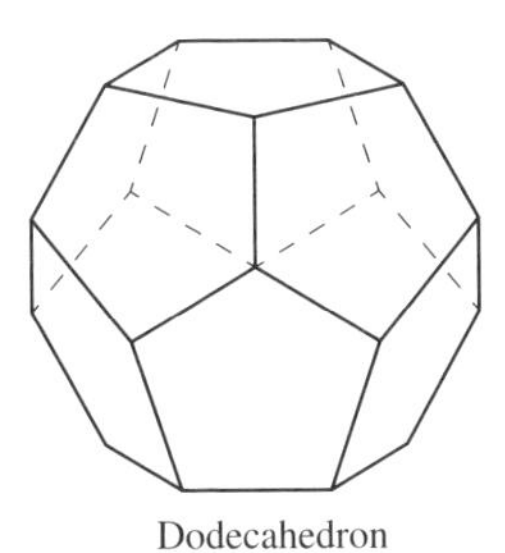

Dodecahedron

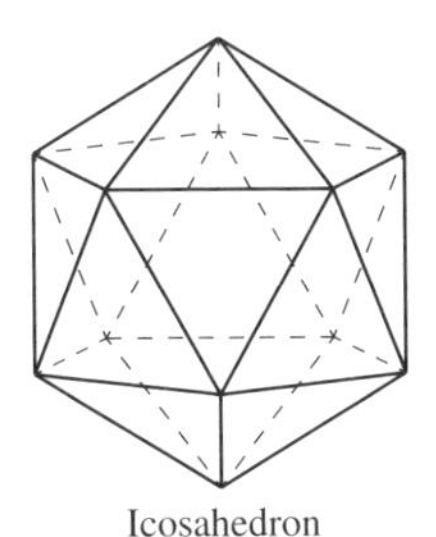

Icosahedron

- **Tetrahedron**—4 faces, which are equilateral triangles. Three triangles meet at each vertex.
- **Hexahedron (Cube)**—6 faces, which are squares. Three squares meet at each vertex.
- **Octahedron**—8 faces, which are equilateral triangles. Four triangles meet at each vertex.
- **Dodecahedron**—12 faces, which are regular pentagons. Three pentagons meet at each vertex.
- **Icosahedron**—20 faces, which are equilateral triangles. Five triangles meet at each vertex.

Why only five? Consider what happens when you connect two or more polygons at a single vertex. If you connect only two polygons, then you wouldn't have a solid. Therefore, to make a polyhedron, you need at least three polygons to meet at each vertex. Additionally, you need the sum of angles at each vertex to be less than 360°, because 360° is a flat plane, not a three-dimensional shape. This greatly limits the types of regular polyhedra that can be formed.

For example, if three squares meet at a single vertex, you may have part of a cube. If four squares meet at a single vertex, you would have a flat plane, not a solid. And, try as you might, you would never be able to connect five or more squares at a single vertex—it is simply impossible.

SPHERES

A **sphere** is a three-dimensional solid that is perfectly round. It is defined as the set of all points in space that are a given distance from its center. A sphere has no edges or vertices and is not a polyhedron.

The distance from the center to the surface of a sphere is called the **radius**, and twice the radius is called the **diameter**.

A sphere is formed by rotating a circle (or semicircle) about its diameter.

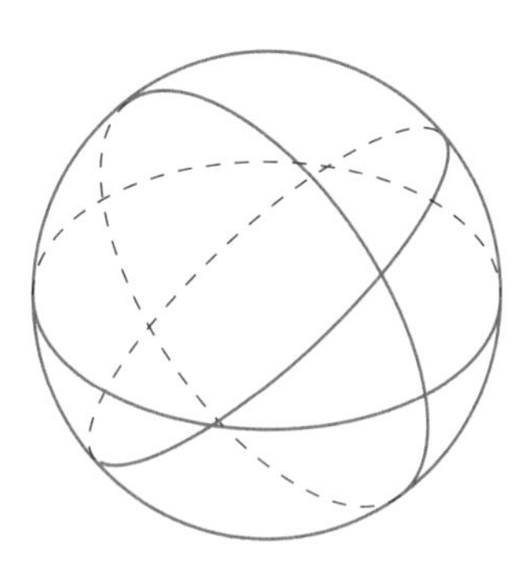

The **cross-section** of a sphere is always a circle. Cross-section is the term for the intersection of a plane through a solid, which forms a two-dimensional shape. Imagine slicing through an orange—the flat surface of the cut piece would be an approximate circle. If you slice through a perfectly spherical solid, the cross section would always be a perfect circle (not an **ellipse**), no matter what angle you cut it from.

A **great circle** is the largest circle that can be drawn around a given sphere. If you make a cross section that passes through the sphere's center, this forms a great circle. If a cross section does not pass through the sphere's center, that's known as a **small circle**.

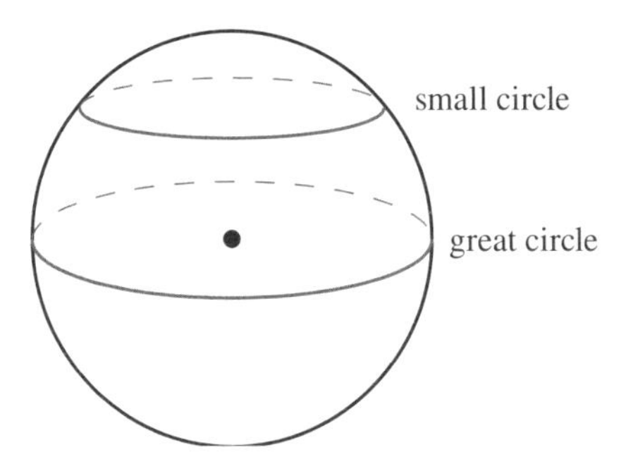

PRISMS

A **prism** is a polyhedron that has a pair of congruent, parallel faces (called the **bases** of the prism). The bases are on opposite ends of the prism, and the other faces (sometimes called **side faces**) are always parallelograms.

A prism is named for the shape of its base—for example, "triangular prism" or "pentagonal prism." In the figures below, the shaded parts show one base of each prism.

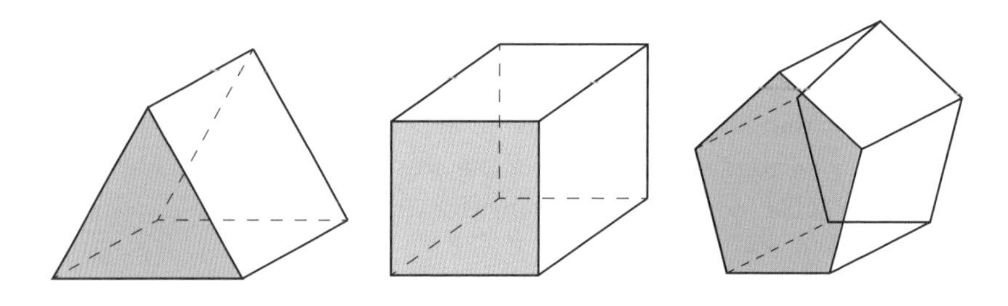

A prism is **regular** if its bases are regular polygons. Otherwise, the prism is **irregular**.

A prism is **right** if the bases are perpendicular to the other faces. Otherwise, it is an **oblique** prism. In a right prism, the side faces are always rectangles. All of the prisms in the above figure are right prisms.

Any cross-section parallel to the base will always be congruent to the base.

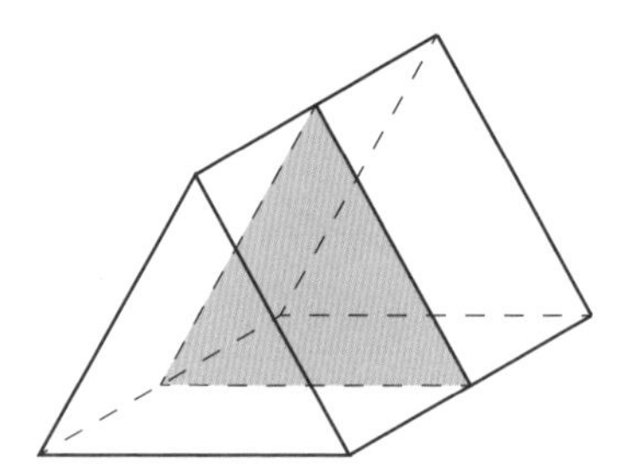

Other cross-sections will often form rectangles or parallelograms, but they may also form triangles, trapezoids, or other shapes. The number of sides in your cross-section is equal to the number of faces you slice through!

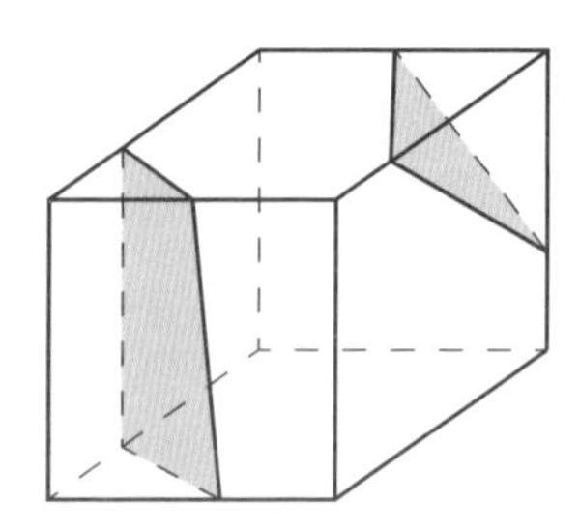

CYLINDERS

A **cylinder** is analogous to a prism, but its bases are circles. The **radius** of a cylinder is the radius of its circular base, and twice the radius is the **diameter**. The distance between the two bases is referred to as the **height**.

A cylinder is formed by rotating a rectangle about its edge, or about its center line.

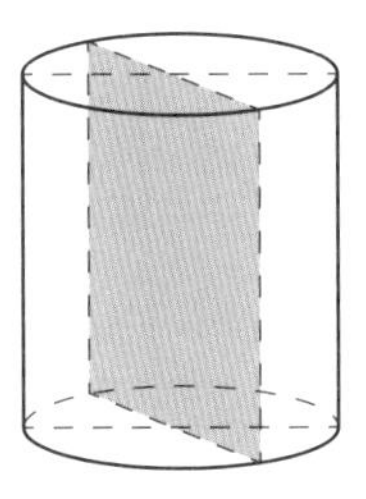

Any cross-section parallel to the base will be a circle congruent to the base.

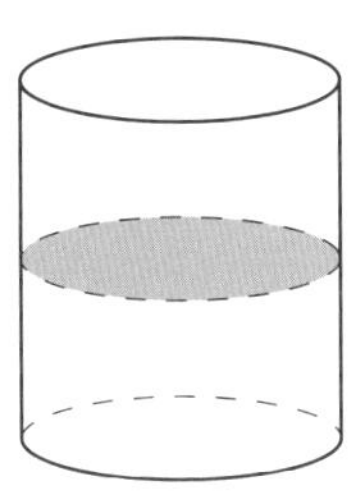

A cross-section perpendicular to the base will always be a rectangle.

"Slanted" cross-sections will be an **ellipse**, or a truncated ellipse (an ellipse with one or both ends cut off).

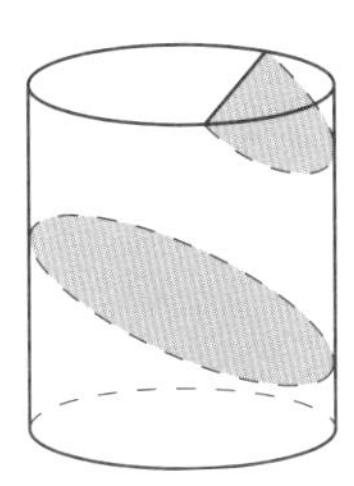

PYRAMIDS

A **pyramid** is a solid with a polygonal base, and triangular faces that meet at a vertex. This "top" vertex of the pyramid is known as the **apex**. The **height** or **altitude** of the pyramid is a line drawn from the apex perpendicular to the base. The **slant height**, conversely, is measured along a two-dimensional face.

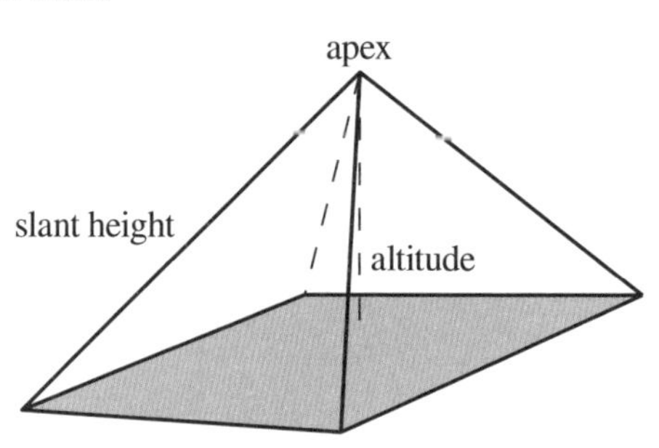

A pyramid is named for the shape of its base—for example, "triangular pyramid" or "pentagonal pyramid."

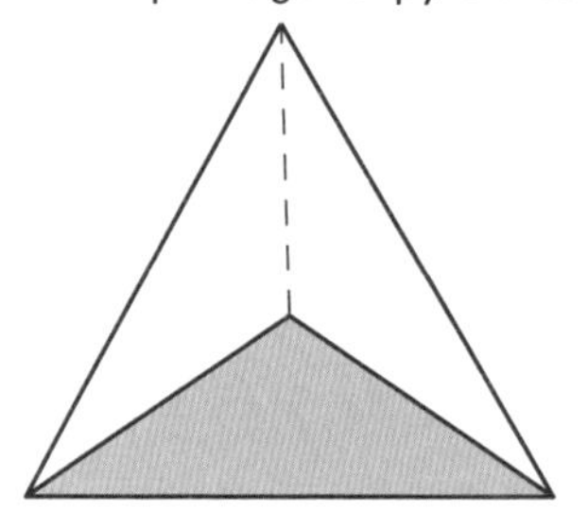

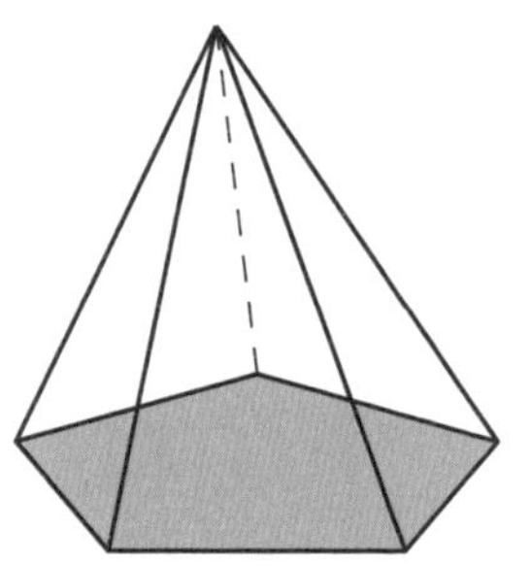

A pyramid is **regular** if its base is a regular polygon. Otherwise, the pyramid is **irregular**.

A pyramid is **right** if the apex is directly above the center of the base. Otherwise, it is an **oblique** pyramid.

In a right pyramid, any cross-section parallel to the base will be **similar** to the base, but smaller. The cross-section is **not** congruent to the base.

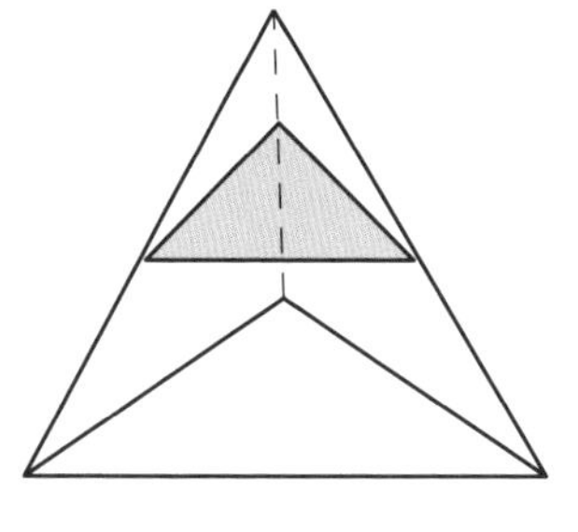

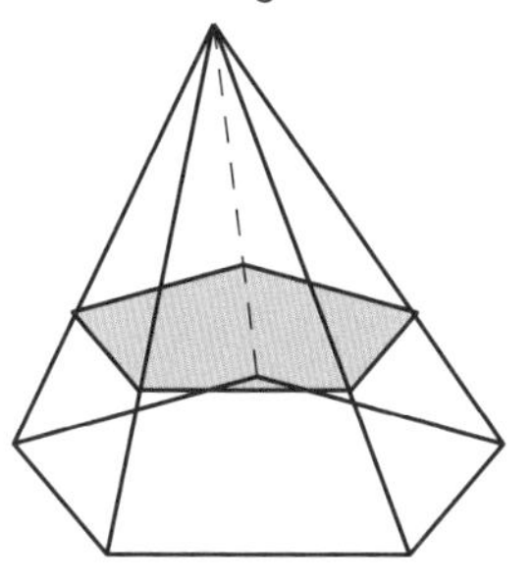

Other cross-sections may also form triangles, quadrilaterals, or other shapes.

CONES

A **cone** is analogous to a pyramid, but it has a circular base. Other than the base, a cone is considered to have one "side," which is curved. The top point of a cone is called the **apex**.

A cone is formed by rotating a right triangle about one of its legs. The **radius** of the base of the cone would be equal to the other leg of the triangle.

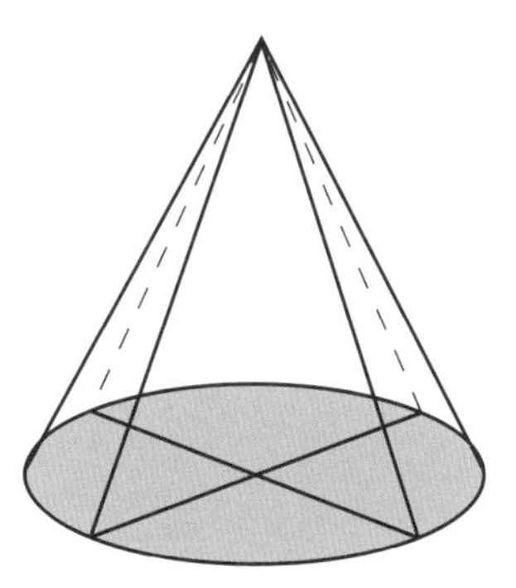

Any cross-section parallel to the base will be a circle, which is smaller than the base.

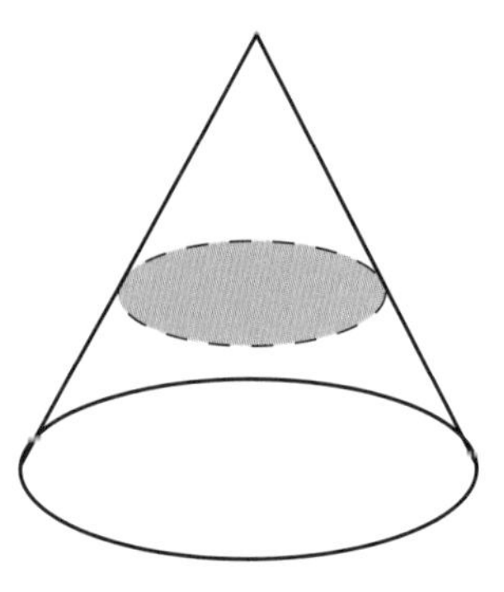

Other cross-sections will be ellipses or parabolas.

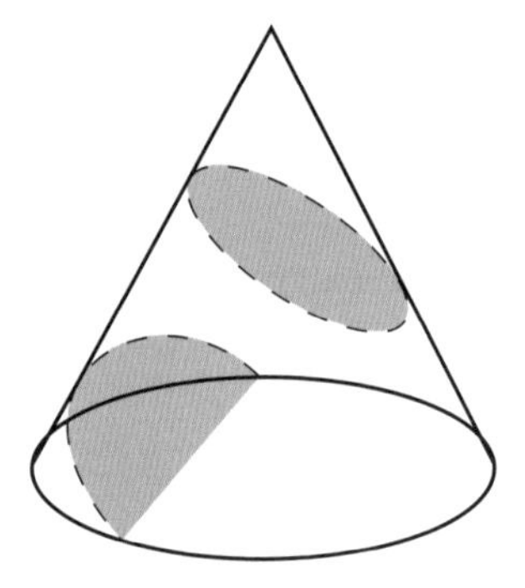

VOLUME AND SURFACE AREA

Volume is the amount of three-dimensional space that is occupied by a solid. You've probably noticed that beverages and other fluids are sold in containers marked for volume—quarts, gallons, and so on. You can also measure volume in cubic units, such as the cubic inch (a cube that is 1 inch on every side). In this chapter, you'll learn volume formulas for some of the most common types of three-dimensional solids. You can use these formulas on more complex figures too, by breaking up a shape into smaller, more recognizable pieces.

Surface area is the amount of two-dimensional area that is taken up by the **surface** of a figure. For example, the surface area of a cube is the sum of the areas of each of its six faces. In this chapter, you'll learn surface area formulas for some of the most common types of three-dimensional solids. To calculate surface area on more complex figures, just remember that you need to calculate the area of every face on the figure.

PRISMS AND CYLINDERS

You can think of a prism as a two-dimensional shape that is stacked on top of itself to have a non-zero height. For example, a single sheet of sticky note paper can be thought of as a two-dimensional rectangle, but a whole pad of sticky notes would be a prism.

VOLUME

Some types of prisms have volume formulas that are fairly easy to remember. But whenever you don't know a formula for a certain prism, here is a general rule that works for every one: to calculate the volume for a prism, first find the area of the **base** (one of the two congruent faces on opposite sides of the figure); then multiply by the **height**.

Volume of a Prism

The volume of any prism is equal to the **area** of the base, multiplied by the height of the prism.

Volume = $B \times h$

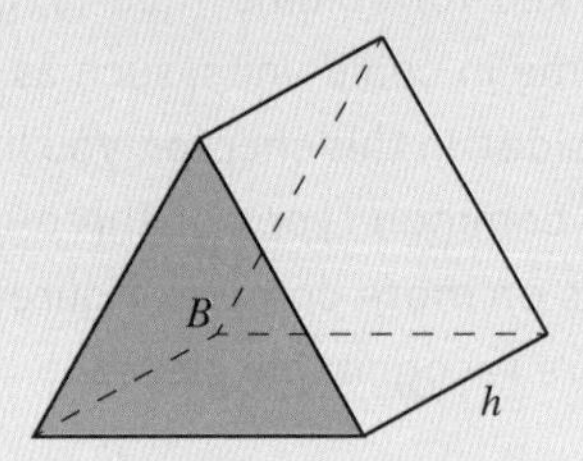

To calculate the volume of a rectangular prism, you can use this straightforward formula:

Volume of a Rectangular Prism

Where l is the length, w is the width, and h is the height of the prism:

Volume = $l \times w \times h$

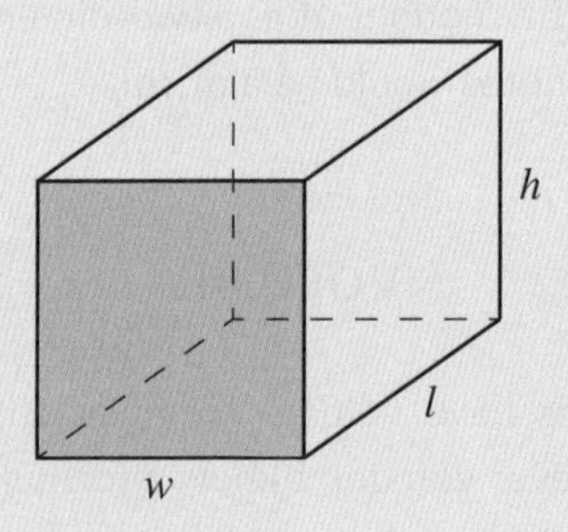

For a triangular prism, we can derive the following formula:

Volume of a Triangular Prism

Where b is the length of the triangular base, a is the altitude of the triangular base, and h is the height of the prism:

$$\text{Volume} = \frac{1}{2} \times a \times b \times h$$

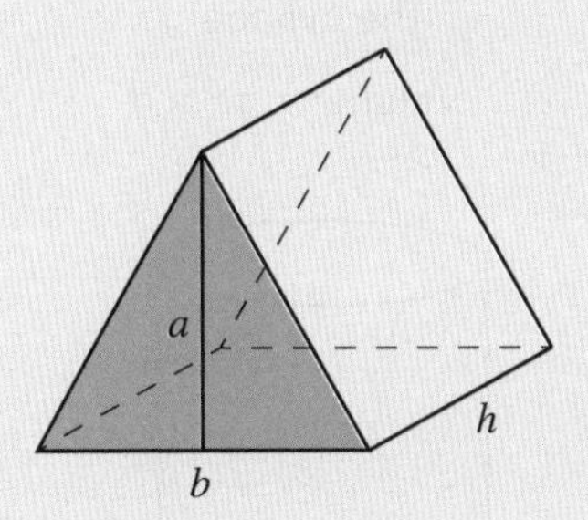

VOLUME OF A CYLINDER

The volume of a cylinder is closely related to the volume of a prism. Using the formula for a prism, we can derive the following formula:

Volume of a Cylinder

Where r is the radius of the circular base, and h is the height of the cylinder:

$$\text{Volume} = \pi r^2 \times h$$

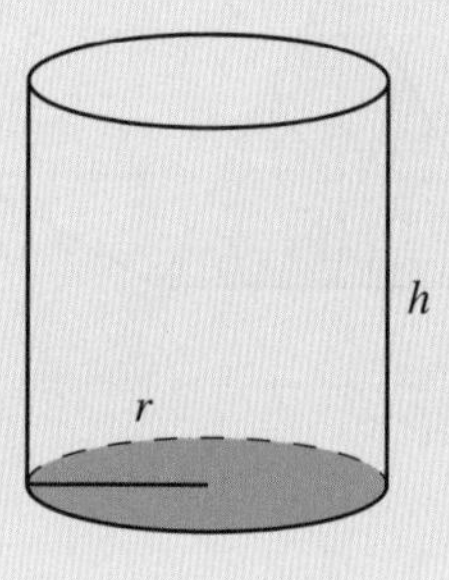

SURFACE AREA

As with volume, some types of prisms have formulas for surface area that are fairly straightforward. But whenever you don't know a formula for a certain prism, the general rule is to add up the areas of every face.

Surface Area of a Prism

The surface area (*SA*) of any prism is the sum of the areas of all of its faces.

Surface Area of a Rectangular Prism

Where l is the length, w is the width, and h is the height of the prism:

Surface Area = $2(lw + lh + wh)$

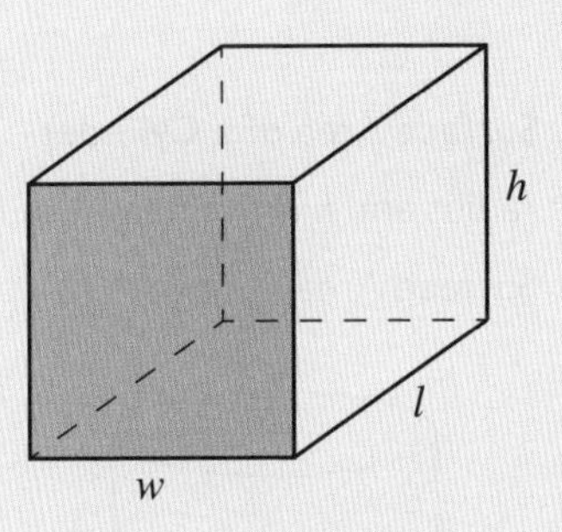

Surface Area of a Prism with a Right Triangle Base

Where a and b are the legs of the triangle, c is the hypotenuse of the triangle, and h is the height of the prism:

Surface Area = $ab + ah + bh + ch$

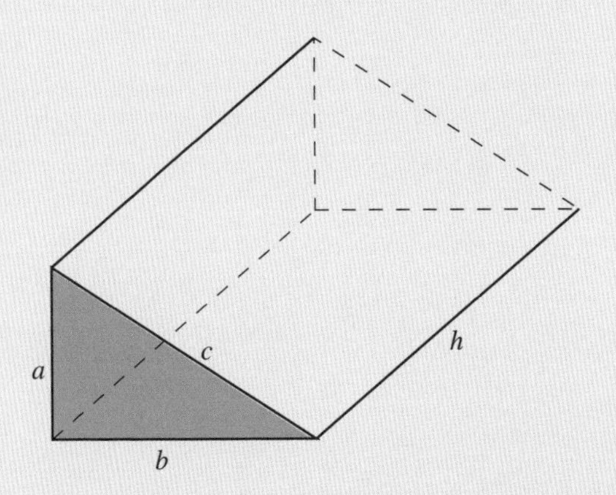

SURFACE AREA OF A CYLINDER

The surface area of a cylinder is closely related to the surface area of a prism. Using what we know about prisms, we can derive the following formula:

Surface Area of a Cylinder

Where r is the radius and h is the height of the cylinder:

$$\text{Surface Area} = 2\pi(r^2 + rh)$$

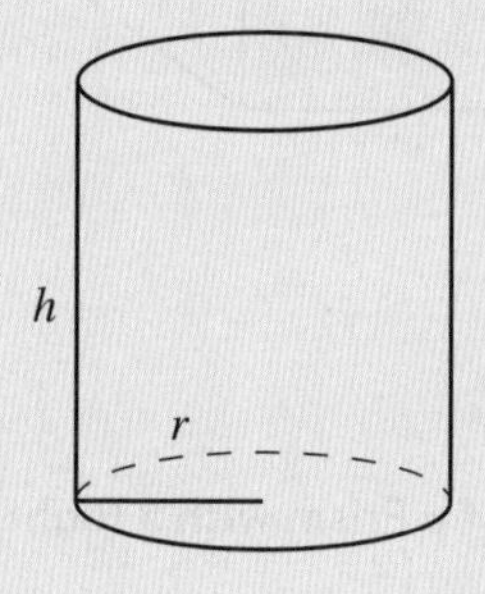

PYRAMIDS AND CONES

VOLUME

Once you are comfortable with finding the volume of a prism, finding the volume of a pyramid is very simple: calculate the volume as though the figure is a prism; then divide by 3.

Volume of a Pyramid

The volume of any pyramid is $\frac{1}{3}$ the **area** of the base, multiplied by the height of the pyramid.

$$\text{Volume} = \frac{1}{3} \times B \times h$$

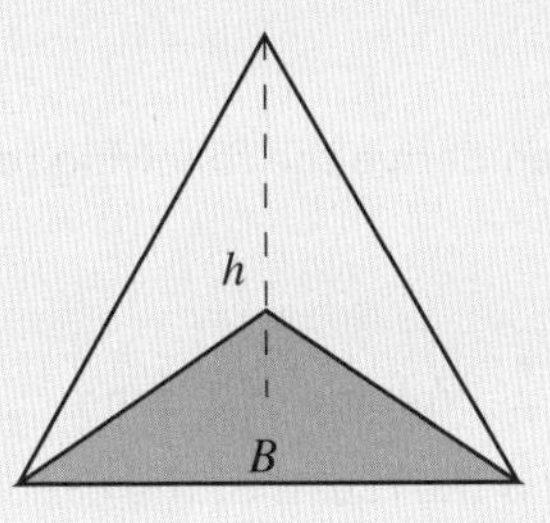

We can illustrate this relationship by combining three congruent pyramids to form a cube.

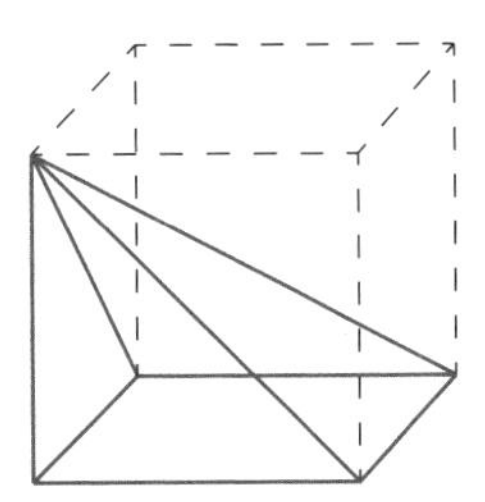

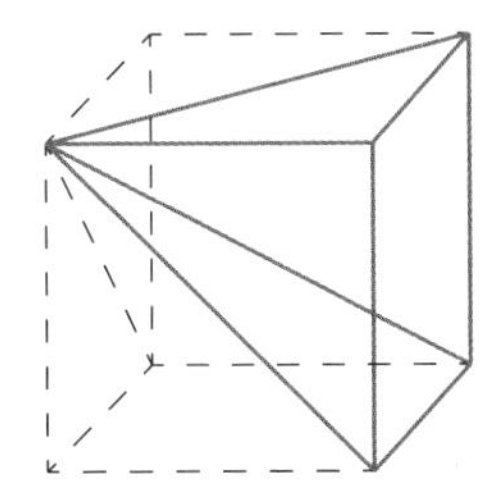

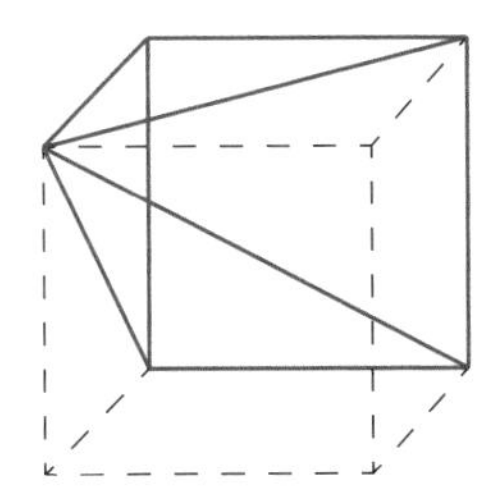

To calculate the volume of a rectangular pyramid, you can use this formula:

Volume of a Rectangular Pyramid

Where l is the length, w is the width, and h is the height of the pyramid:

$$\text{Volume} = \frac{1}{3} \times l \times w \times h$$

VOLUME OF A CONE

The process for calculating the volume of a cone is similar to that for a pyramid: calculate the volume as though the figure is a cylinder; then divide by 3.

Using these rules, we can derive the following formula:

Volume of a Cone

Where r is the radius of the circular base, and h is the height of the cone:

$$\text{Volume} = \frac{1}{3} \times \pi r^2 \times h$$

SURFACE AREA

To calculate the surface area of a pyramid, the most straightforward thing to do is to calculate the area of every face.

Surface Area of a Pyramid

The surface area of any pyramid is the sum of the areas of all of its faces.

SURFACE AREA OF A CONE

Surface Area of a Cone

Where r is the radius and s is the slant height of the cone:

$$SA = \pi r^2 + \pi rs$$

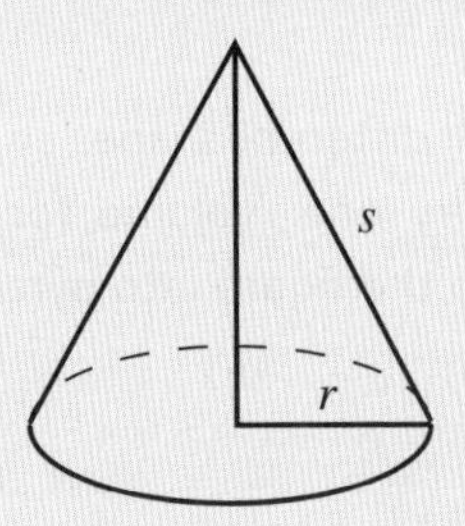

The formula for surface area of a cone is derived from the relationship between two circles. If you cut a cone open and flatten it out, you would have a portion of a circle:

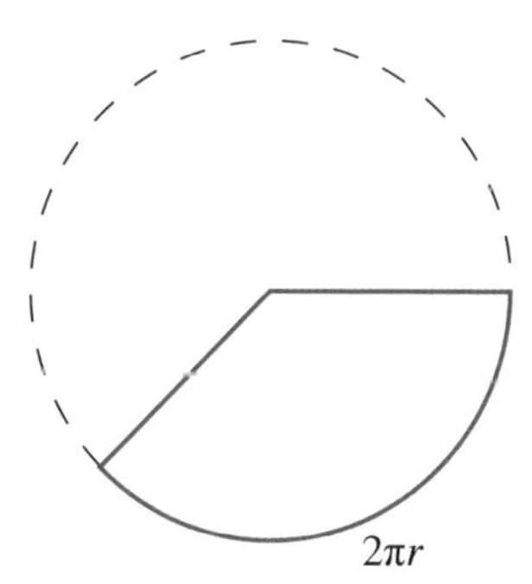

If we know the radius and arc length of a portion of a circle, then we can calculate the area of the sector of the circle.

POLYHEDRA

Since polyhedra can be all different shapes and sizes, there is no simple formula to apply to every one. Volume for a polyhedron is usually calculated by breaking up the figure into pyramids, and calculating the volume of each pyramid.

SPHERES

Volume of a Sphere

Where r is the radius:

$$V = \frac{4}{3}\pi r^3$$

Surface Area of a Sphere

Where r is the radius:

$$A = 4\pi r^2$$

The Greek philosopher Archimedes discovered this formula over two thousand years ago, when he theorized that the surface area of a sphere is equal to that of its inscribed cylinder.

CHAPTER

11

REASONING AND PROOFS

Proofs are at the heart of mathematics. Your teachers ask you to show your work, because how you arrive at the answer is more important than the answer itself. Proofs are the reason the laws of mathematics can be stated with certainty.

CHAPTER CONTENTS

INDUCTIVE AND DEDUCTIVE REASONING

INDUCTIVE REASONING

Inductive reasoning is reasoning based on patterns. When you use observations or examples to form a conclusion, you are using inductive reasoning. A conclusion reached from inductive reasoning is called a **conjecture**.

Inductive reasoning is useful for forming ideas and hypotheses, which then require further research. However, it cannot be used to fully *prove* that something is true.

For example, a person might observe that they feel sick whenever they eat microwave popcorn. They might then conclude that they are allergic or intolerant to popcorn, and they might be correct. However, there are other possible explanations that would need to be considered, such as the flavoring ingredients in the popcorn they have eaten. Therefore, the original conclusion might not be true for all types of popcorn.

In mathematics, we use inductive reasoning when we work with patterns. For example, if we observe a town's population for three years, and see that it increases from 1,000 to 1,500 and then to 2,000, we might conclude that the population will continue to increase by 500 per year. This is reasoning based on an observed pattern.

Nearly all of the mathematical rules we know, such as the rules of geometry we reviewed in this book, started as conjectures. For example, before discovering the constant number π, mathematicians may have observed that they always seemed to get the same value when dividing any circle's circumference by its diameter. They would have continued investigating why this was true until it was fully proved.

You can prove that a conjecture is false by providing a single **counterexample**. A counterexample is simply an example for which the conjecture is not true. For instance, if someone said *all quadrilaterals have two congruent diagonals*, you would just need to show *one* example of a quadrilateral that does not have this property. That would prove that the conjecture is false, since there is at least one case for which it is false.

Conjecture: All quadrilaterals have two congruent diagonals.

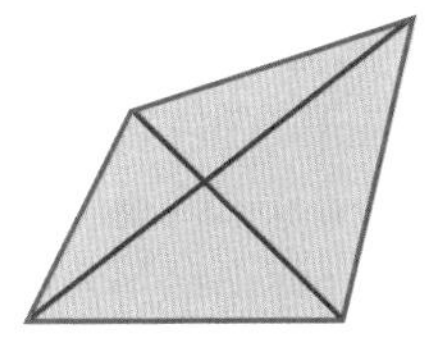

Counterexample: This quadrilateral does ***not*** have two congruent diagonals.

DEDUCTIVE REASONING

Deductive reasoning is reasoning based on known facts.

In geometry, when we use established rules to find missing information about a figure, we are using deductive reasoning. For example, consider a triangle with two known angle measures:

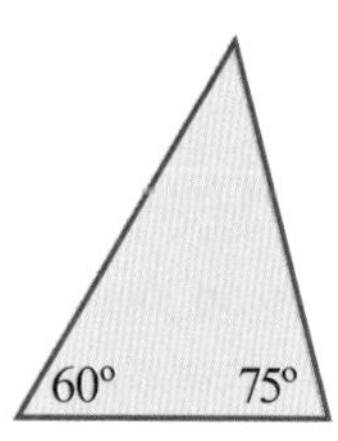

We can use deductive reasoning to find more information:

1. Fact: the figure is a triangle.
2. Fact: for any triangle, the sum of angle measures is 180°.
3. Therefore, for this triangle, the sum of angle measures is equal to 180°.
4. The third angle measure is $180° - 60° - 75° = 45°$.

We also use deductive reasoning whenever we solve an equation. With every step, we are using an established rule that is known to be true. For example, if we subtract a number from both sides of an equation, we are using the *subtraction property of equality*, so we know that the equation remains true after this step.

You can think of the difference this way: inductive reasoning uses examples to try to make a rule, while deductive reasoning uses rules to apply to an example.

CONDITIONAL STATEMENTS

A **conditional statement** is also known as an **if-then statement**. In other words, it is a statement using the form "If ___ is true, then ___ must also be true." For example, we could say, "If we are at Disneyland, then we are in California." This is a true conditional statement, since Disneyland is in California. A conditional statement can either be *true* or *false*. This is known as the **truth value** of the statement. The statement is true if it is *always* true, and it is false if it is *not always* true.

In a conditional statement, the part with *if* is called the **hypothesis** of the statement, and the part with *then* is its **conclusion**. In the previous example, the hypothesis of the statement is *if we are at Disneyland*, and the conclusion is *then we are in California*.

Conditional statements can be written with symbols. Most commonly, you will see $\boldsymbol{p \rightarrow q}$. We read this notation as *if p, then q*. That is, *p* is the hypothesis, and *q* is the conclusion.

Conditional Statement

$$\boldsymbol{p \rightarrow q}$$

If ***p***, then ***q***

RELATED STATEMENTS

Whenever you make a conditional statement, there are several other related statements with different meanings.

The **negation** of a statement is its opposite, usually using the word *not*. For example, the negation of "we are in California" is "we are *not* in California." Using symbols, we write $\sim p$ to mean "not *p*," or, "*p* is not true."

Negation Statement

$\sim p$

Not p

The **converse** of a statement switches the original statement's hypothesis and conclusion. For example, for the statement *If we are in Disneyland, then we are in California*, the converse is *If we are in California, then we are in Disneyland.*

Note that the converse of a statement does **not** necessarily have the same truth value of the original statement. In our example, the converse statement is not true, because, of course, we might be in California without being in Disneyland.

Converse Statement

$q \rightarrow p$

If ***q***, then ***p*** is the converse of if ***p***, then ***q***.

The **inverse** of a statement is the negation of both the hypothesis and the conclusion. For our example, the inverse of the original statement is *if we are not in Disneyland, then we are not in California.*

The inverse of a statement does **not** necessarily have the same truth value of the original statement.

Inverse Statement

$\sim p \rightarrow \sim q$

If ***not p***, then ***not q*** is the inverse of if ***p***, then ***q*.**

The **contrapositive** of a statement switches *and* negates the original hypothesis and conclusion. For our example, the contrapositive of the original statement is *if we are not in California, then we are not in Disneyland.* Note that this is true because Disneyland is in California.

The contrapositive of a statement *does* have the same truth value of the original statement.

Contrapositive Statement

$\sim q \rightarrow \sim p$

If ***not q***, then ***not p*** is the contrapositive of *if **p**, then **q**.*

Equivalent statements are statements that have the same truth value. Given a conditional statement, the contrapositive is equivalent to the original statement. In other words, they are either both true or both false. Also, the original statement's converse and inverse are equivalent to each other!

Sometimes, a conditional statement and its converse are both true. When this is the case, we can write a **biconditional statement** to combine the statement and its converse. A biconditional statement usually has the phrase *if and only if.*

For example, consider the statement *a rectangle has four right angles.* This is true—all rectangles have four right angles. It is also a kind of conditional statement. We could even rewrite it by saying *if a quadrilateral is a rectangle, then it has four right angles.*

Now consider the converse of the statement: *if a quadrilateral has four right angles, then it is a rectangle.* This is also true—all quadrilaterals that have four right angles are rectangles.

Since the statement and converse are both true, we can combine them with a biconditional statement: *a quadrilateral is a rectangle if and only if it has four right angles.* In this case, we can also switch the two statements—*a quadrilateral has four right angles if and only if it is a rectangle.*

For these reasons, biconditional statements are often used to *define* terms and concepts in mathematics. A biconditional makes a definition clear, because it tells us concisely that the statement and its converse are both true. That is to say, the phrase *if and only if* is meaningful and important in mathematics.

Biconditional Statement

$$p \leftrightarrow q$$

p is true if and only if q is true.

PROOFS

In geometry, we often have to use limited information about a problem in order to figure out several unknowns. Mathematicians call it a **proof** when they use known information to derive other facts and unknowns in a problem. In fact, many geometry classes spend a great deal of time writing exhaustively detailed proofs for problems and figures. We won't make you write full proofs in this book! However, in this section we will review some basics.

Throughout history, mathematicians have developed several rules that we now use to solve problems. The words **theorem** and **postulate** both describe rules. These terms are often used interchangeably, and you probably shouldn't worry about the differences. But just so you know: a postulate (also known as an **axiom**) is something that everyone agrees is true, while a theorem is something that needs to be proven true using logical steps. Many of the "theorems" that we discuss today are effectively postulates, because they've been proven true in the past and we no longer need to doubt their accuracy.

In mathematics, a **proof** is a set of logical steps that one can use to show that something is true. That "something" may be a theorem, postulate, definition, or some other property of a figure or problem. The reason that proofs are such a popular teaching topic is that they help students gain an intuitive understanding of why things work the way that they do.

In this first example, we'll develop not a formal proof, but an explanation for a known postulate.

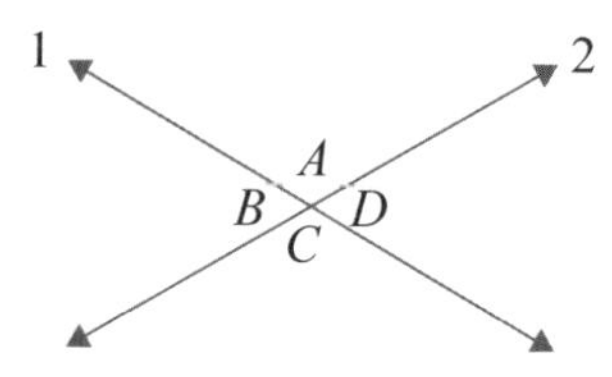

Given that 1 and 2 are lines, explain why $\angle \mathbf{A} = \angle \mathbf{C}$.

The real question here is why vertical angles are always congruent. You have probably learned the postulate that states that "vertical angles are always congruent," but since we're trying to explain the concept, we're not just going to cite the postulate and call it done. Instead, we'll find reasons to support our explanation.

First, let's consider everything we know about this figure.

Of course, we know that both 1 and 2 are straight lines, which is a fact given in the problem. That's important, because we know that straight lines are 180°. That means that each pair of adjacent angles in the figure has a sum of 180°. When two angles form a line as in these examples, they are called a **linear pair**.

Therefore,

$$\angle A + \angle B = 180°$$

$$\angle B + \angle C = 180°$$

$$\angle C + \angle D = 180°$$

$$\angle D + \angle A = 180°$$

If you look at the list of pairs above, you'll notice equations that have an angle in common. For example, ($\angle A + \angle B = 180°$) and ($\angle B + \angle C = 180°$) both refer to $\angle B$. This allows us to use algebra skills to make some deductions.

What if we subtract $\angle B$ in both of those equations? Note: to make this easier to read, we'll refer to the angles by their letters, and omit the angle symbol.

$A + B = 180°$

$A + B - B = 180° - B$ Subtract B from both sides.

$A = 180° - B$

$B + C = 180°$

$B - B + C = 180° - B$ Subtract B from both sides.

$C = 180° - B$

Therefore, we've proved that both *A* and *C* equal the quantity (180° – *B*).

In the following example, you'll encounter a formal proof.

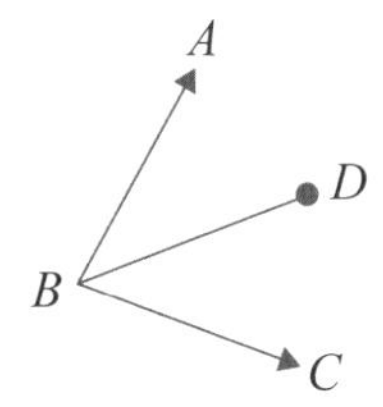

In the figure above, the measure of ∠***ABC*** is 80°, and ∠***ABC*** is bisected by line segment ***BD***. What is the measure of ∠***ABD***?

If you were asked to explain your answer to this problem, you'd probably say something like this:

"If segment *BD* bisects angle *ABC*, that means it divides the angle exactly in half. Therefore, each of the two halves must equal 40°. The measure of ∠*ABD* is 40°."

If you said that, you'd be correct! However, if you were asked to *write a proof* for this problem, then your explanation should be different. Here's how a proof would look for this example.

You'll often find that there are many possible ways to support a particular proof. That's normal! There's rarely just one "right" answer when it comes to proofs.

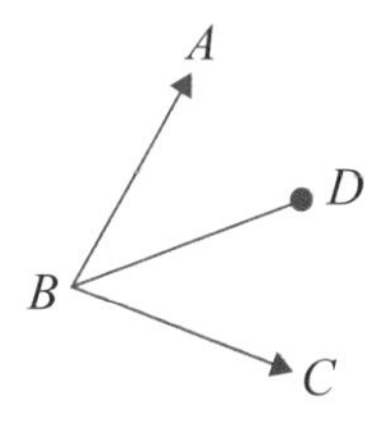

Given: $\angle ABC = 80°$
BD bisects $\angle ABC$

Prove: $\angle ABD = 40°$

Notice the format of this problem. In a proof exercise, you'll often see included information labeled **given**, and a conclusion statement labeled **prove**. Your job is to document the logical steps that lead from the given information to the conclusion.

In writing your proof, it's conventional to use two columns, with "Statements" on the left and "Reasons" on the right. You'll usually begin by including some or all of the given information in your first statement(s). Also, if no figure has been provided, you should always draw one yourself, and label all of the given information in your figure.

Statements	Reasons
1. $\angle ABC = 80°$ BD bisects $\angle ABC$	1. Given

Then, you would continue the proof by writing statements that can be derived from the given information. For each statement on the left, you must write your *reason* or justification for that statement on the right. Reasons are always general theorems or postulates—don't use specific names from your figure in the "reasons" column.

Here's how a complete proof might look for this example:

Statements	Reasons
1. $\angle ABC = 80°$ BD bisects $\angle ABC$	1. Given
2. $\angle ABD = \angle ACD$	2. Angle bisector divides an angle into two equal parts
3. $\angle ABD = \angle ABC \div 2$	3. Division produces equal parts
4. $\angle ABD = 80° \div 2$	4. Substitution
5. $\angle ABD = 40°$	5. Division

In a proof, the order of your statements can vary, as long as every statement logically follows from something above it. Additionally, you must never skip steps when writing a formal proof—if a particular fact leads to your next statement, then you must include it in writing, even if it's very obvious. Some instructors are stricter than others when it comes to obvious steps, but it's better to be safe than sorry. Finally, you do not need to include *every* possible fact about the figure in your proof—only those that are necessary and sufficient to reach the conclusion. For example, in this problem, we didn't need to specifically mention that $\angle ACD$ also equals 40°, even though that can be proved as well.

NOTES

NOTES

NOTES